AF789210

ON SCIENCE, INFERENCE, INFORMATION AND DECISION-MAKING

SYNTHESE LIBRARY

STUDIES IN EPISTEMOLOGY,

LOGIC, METHODOLOGY, AND PHILOSOPHY OF SCIENCE

Managing Editor:

JAAKKO HINTIKKA, *Boston University*

Editors:

DIRK VAN DALEN, *University of Utrecht, The Netherlands*
DONALD DAVIDSON, *University of California, Berkeley*
THEO A.F. KUIPERS, *University of Groningen, The Netherlands*
PATRICK SUPPES, *Stanford University, California*
JAN WOLEŃSKI, *Jagiellonian University, Kraków, Poland*

VOLUME 271

Klemens Szaniawski
1925–1990

Photograph by Tadeusz Póżniak

KLEMENS SZANIAWSKI

ON SCIENCE, INFERENCE, INFORMATION AND DECISION-MAKING

Selected Essays in the Philosophy of Science

Edited by

ADAM CHMIELEWSKI
University of Wrocław, Poland

and

JAN WOLEŃSKI
Jagiellonian University, Kraków, Poland

SPRINGER SCIENCE+BUSINESS MEDIA, B.V.

Library of Congress Cataloging-in-Publication Data

A catalogue record for this book is available from the Library of Congress.

ISBN 978-94-010-6213-8 ISBN 978-94-011-5260-0 (eBook)
DOI 10.1007/978-94-011-5260-0

Printed on acid-free paper

All Rights Reserved
© 1998 Springer Science+Business Media Dordrecht
Originally published by Kluwer Academic Publishers in 1998
Softcover reprint of the hardcover 1st edition 1998

No part of the material protected by this copyright notice may be reproduced or utilized in any form or by any means, electronic, mechanical, including photocopying, recording or by any information storage and retrieval system, without written permission from the copyright owner.

TABLE OF CONTENTS

PREFACE

Klemens Szaniawski was born in Warsaw on March 3, 1925. He began to study philosophy in the clandestine Warsaw University during World War II. Tadeusz Kotarbiński, Jan Łukasiewicz, Maria and Stanisław Ossowskis, Władysław Tatarkiewicz, and Henryk Hiż were among his teachers. Szaniawski was also a member of the Polish Home Army (AK), one of the youngest. He was arrested and spent the last period of the war as a prisoner in Auschwitz. After 1945, he continued his studies in the University of Łódź; his Master thesis was devoted to French moral thought of the 17th and 18th centuries. Then he worked in the Department of Ethics in Łódź. In 1950, he received his Ph.D. on the basis of the dissertation on the concept of honour in knight groups in the Middle Ages; Maria Ossowska was the supervisor. In the early fifties he moved to Warsaw to the Department of Logic, directed by Kotarbiński. He took his habilitation exams in 1961. In 1969 he became a professor. Since 1970 he was the head of Department of the Logic at the Warsaw University. In the sixties Szaniawski was also the Dean of the Faculty of Philosophy and Sociology. In 1984 he was elected the Rector Magnificus of the Warsaw University but the Ministry overruled the autonomous democratic vote of the academic community. He served as the President of the Polish Philosophical Association (since 1977) taking this post after Kotarbiński. He was also the Editor-in-chief of *Studia Logica* (1970–1974) and a member of editorial boards of many other journals, *Synthese* included.

Klemens Szaniawski was also a very important political and public figure. He never accepted communism and its ideology. In the late sixties, he became one of the leading Polish dissidents. As the Dean of the Faculty in 1968, he inspired a very strong protest of the Council of the Faculty against antisemitism in Poland. In the seventies, he was one of the organizers of the group called "Experience and Future", which discussed the political situation in Poland and elaborated several warning reports on the decline of Polish social life. When "Solidarity" movement came into being in 1980, Szaniawski immediately joined its ranks and soon became one of its most important members. As a great master of dialogue and compromise, he successfully participated in several negotiations between "Solidarity" and the government. He also organized a special association of Polish artistic and scientific societies responsible for independent cultural life in Poland. He was the chief person responsible for the organization of the Congress of Polish Culture which was to be held in the middle of December 1981. The Martial Law introduced in

Poland on December 13, 1981 prevented this event. Immediately after introduction of the Martial Law, Szaniawski was detained, but he was released after two days. He immediately became a part of various underground activities. In particular, he was the head of the special committee which organized financial support for repressed scientists and artists in Poland. He also served as one of the closest political advisors of Lech Wałęsa. When the so-called Round Table was prepared in Poland in 1989, Wałęsa asked Szaniawski to join the official representation delegation of representatives of "Solidarity" for negotiations with the government. I well remember TV news on the day when the Round Table started. When the "Solidarity" group led by Wałęsa entered the building, Szaniawski was the fourth from the top. It shows his actually high position in "Solidarity". However, he faced a very difficult dilemma at the time. After seven years of a strict prohibition of going abroad, Szaniawski had just received a permission and wanted to go for a scholarship in the USA, which had waited for him since the early eighties (it is quite possible that the government intended to prevent his participation in the Round Table). Since any further prolongation was absolutely impossible, Szaniawski decided to resign from his place at the Round Table and informed Wałęsa about his decision, but the leader of "Solidarity" insisted that he should stay with the entire group at least for one more day which he did. In the end of 1989 he was nominated the Polish ambassador in the United States of America. He contracted cancer in December 1989 and died on March 5, 1990.

Klemens Szaniawski was a charming person. Everybody who knew him was greatly impressed by his personality, a unique sense of humour, openness in personal contacts and his way of thinking that combined refined scepticism with a very firm belief in human rationality. In this he resembled Russell, who was one of Szaniawski's main intellectual heroes.

I met Szaniawski for the first time in July 1957 during summer holidays. At that time, I was in the secondary school and I did not think that I would become a philosopher myself. When I began my undergraduate studies and philosophy of science became one of my favourites, I very often read his exceptionally clear writings; I learned very much from them. Then, we met occasionally on various meetings, but these encounters were rather formal, although he well remembered that we had met earlier. In the seventies, our contacts were more frequent, because we both participated in regular meetings in philosophy of sciences. Once he said to me: "Jan, I do not understand why we still address each other so formally. I am Klemens." I was very proud that such a distinguished academic and political figure offered me this cordial relationship. In the eighties, we met very often. I always visited him in Warsaw and he did the same when he came to Wrocław, where I lived at the time. He enjoyed very much eating red borsch prepared by my wife. Once he said: "Oh, I must come to you every month to taste this excellent soup."

Klemens was an extraordinary authority in all problems. I remember the following event. The Austrian Cultural Institute in Warsaw planned to organize an international seminar on Austro-Polish philosophical relationships. Mr. Sickinger, the head of the Institute invited a few Polish scholars, including Klemens and myself, in order to discuss this project. I was asked to prepare a list of participants and a preliminary program. Toward the end of an excellent dinner, I made some notes in my notebook. After returning to Wrocław, I found a very elegant gold "Parker" ball-point in my pocket. I immediately realized that it belonged to Mr. Sickinger who had lent it to me for making notes and which I automatically put in my pocket. I was very desperate and I immediately called Szaniawski for advice. I started with something like this: "Klemens, something terrible happened", and then I reported the story. He answered: "Only this. At first, I thought that really something terrible happened. Do not worry. Call the Austrian Institute and tell them what happened. Mr. Sickinger probably did not realize that you took his ball-point." I followed his advice; Klemens was right: the secretary to Mr. Sickinger was very surprised by my phone call and the whole problem. Let me remark that it was 1984 and much more terrible things happened in Poland at that time; hence, Klemens' answer was "I thought that really something terrible happened."

Szaniawski left Poland in April 1989 and returned in October of the same year. I went to the USA in September 1989 and spent three months there. Once more I visited the USA in January and February of 1990. My contacts with Klemens were limited only to talks on the phone. In September 1989, I called him in Washington, D.C., and we had a long conversation. He told me about his scientific plans which he began to implement during his visit in America. Our last contact took place, also on the phone, in December. He told me: "I am not quite well, I got only a little cold." Nobody expected that it was the beginning of the mortal illness.

Szaniawski's first philosophical interests concerned ethics. However, the period of the early fifties in Poland was not conducive for an independent research in this field. Szaniawski decided to move to formal logic and philosophy of science. When Kazimierz Ajdukiewicz came to Warsaw in 1955, Szaniawski began to work with him. Their common scientific interest concerned fallible inferences. Both intended to build a theory of such reasonings based on results of decision theory and mathematical statistics. Ajdukiewicz's conception of the so-called pragmatic methodology which takes inferential procedures as closely connected with human epistemic activities, became a general background of this research.

Szaniawski was educated in the tradition of the Lvov-Warsaw Philosophical School for which the idea of logical inference played the main role in rational procedures. On the other hand, he was also very strongly influenced by ideas coming from modern statistics. Now, as it is widely known, many statisticians doubt whether the classical concept of inference has any sound application in statistical procedures and their analysis. For example, they think that we have to do with inductive behaviour rather than induction as an inference. Szaniawski wanted to find a compromise between various competing general proposals in the foundations of statistics, probability and decision theory. In particular, he believed that there was a possibility of reconciling Bayesian and non-Bayesian ideas on the one side, and objectivistic and subjectivistic on the other. He observed that the problem of rationality was a common denominator of conflicting ideas in the foundations of inductive inference. Assuming this key idea, Szaniawski attempted to demonstrate that an opposition between inferential and behaviouristic account of induction was fairly apparent. In the sixties, he began to work on an approach according to which science was an information-searching process. During his last visit in the USA, he collected a lot of material for a large monograph on this topic, unfortunately, however, this project was interrupted by his early death. Szaniawski also worked on more specific problems in statistics. In particular he introduced an original b-criterion of rational decision-making, and he generalized the method of sequential tests.

Szaniawski always thought about applications of formal methods in social sciences and ethics. He believed that formal models, for example of distributive justice or distributing of goods, realized basic principles of rationality. In fact, rationalism, understood as anti-irrationalism, was the guiding idea of his philosophy. The maxim *plus ratio quam vis* was particularly important for him. He devoted his life to realize this motto everywhere, also in the social practice. The papers collected in this volume give a picture of Klemens Szaniawski's philosophical thought which is a continuation of the best Polish intellectual tradition. Let this book be also a tribute to an exceptional human being, a teacher, friend and colleague.

Jan Woleński

ACKNOWLEDGEMENTS

The editors and the publisher are grateful to the following persons and institutions for permission to reprint the papers included in this volume:

1. "Some Remarks on the Philosophy of Science" (first published in *Zagadnienia Naukoznawstwa* 2 (1968), 26–32, Ossolineum, Wrocław).
2. "Information and Decision as Tools of Philosophy of Science" (first published in *Danish Yearbook of Philosophy* 10 (1973), 47–59, now published by Museum Tusculanum Press, Copenhagen).
3. "Methods versus Creative Activity in Science" (first published in *Zagadnienia Naukoznawstwa* (special issue), 1974, 5–13, Ossolineum, Wrocław).
4. "Sociology and Models of Rational Behaviour" (first published in *Dialectics and Humanism* 2 (1974), 133–144, Państwowe Wydawnictwo Naukowe, now Wydawnictwo Naukowe – PWN, Warszawa).
5. "Mathematical Models and Social Facts" (first published in *The Polish Sociological Bulletin* 2 (1970), 103–106, Ossolineum, Wrocław).
6. "Science as an Information-seeking Process" (first published in *Postępy Cybernetyki* 3 (1981), 23–31, Ossolineum, Wrocław).
7. "Inference or Behaviour?" (first published in *Studia Filozoficzne* (translations) 1 (1962), 171–179, Państwowe Wydawnictwo Naukowe, now Wydawnictwo Naukowe – PWN, Warszawa).
8. "A Note on Confirmation of Statistical Hypotheses" (first published in *Studia Logica* X (1960), 111–118, Ossolineum, Wrocław).
9. "On Some Basic Patterns of Statistical Inference" (first published in *Studia Logica* XI (1961), 77–87, Ossolineum, Wrocław).
10. "A Method of Deciding between N Statistical Hypotheses" (first published in *Studia Logica* XII (1961), 135–141, Ossolineum, Wrocław).
11. "A Pragmatic Justification of Rules of Statistical Inference" (first published in *The Foundations of Statements and Decisions*, ed. by K. Ajdukiewicz, Państwowe Wydawnictwo Naukowe, now Wydawnictwo Naukowe – PWN), Warszawa 1965, 301–310).
12. "On Sequential Inference" (first published in *Local Induction*, ed. by R. Bogdan, D. Reidel, Dordrecht 1976, 171–182).
13. "Interpretations of the Maximum Likehood Principle" (first published in *Twenty-Five Years of Logical Methodology in Poland*, ed. by M. Przełęcki and R. Wójcicki, Państwowe Wydawnictwo Naukowe, now Wydawnictwo Naukowe – PWN, Warszawa – D. Reidel, Dordrecht 1977, 625–634).
14. "Some Remarks Concerning the Criterion of Rational Decision-Making"

(first published in *Studia Logica* IX (1960), 221–235, Ossolineum, Wrocław).

15. "The Concept of Distribution of Goods" (first published in *Studia Filozoficzne* (translations) 4 (1970), 255–266, Państwowe Wydawnictwo Naukowe, now Wydawnictwo Naukowe – PWN, Warszawa).
16. "The Value of Perfect Information" (first published in *Synthese* 17 (1967), 408–424).
17. "Questions and Their Pragmatic Value" (first published in *Logic, Language and Probability*, ed. by R. Bogdan, D. Reidel, Dordrecht 1973, 121–123).
18. "Two Concepts of Information" (first published in *Theory and Decision* June 1974, v. 5, no. 1, 9–21).
19. "Types of Information and Their Role in the Methodology of Science" (first published in *Formal Methods in the Methodology of Empirical Sciences*, ed. by M. Przełęcki, K. Szaniawski and R. Wójcicki, Ossolineum, Wrocław – D. Reidel, Dordrecht 1976, 297–308).
20. "Information in Decision-Making. Some Logical Aspects" (first published in *Problems in Formalization in the Social Sciences*, ed. by K. Szaniawski, Ossolineum, Wrocław 1977, 77–90).
21. "Decision-Making and Future Research. Some Theoretical Problems" (first published in *Polska 2000. Future Research – Planning – Decision Making*, Ossolineum, Wrocław 1978, 77–90).
22. "On Formal Aspects of Distributive Justice" (first published in *Essays in Honour of Jaakko Hintikka on the Occasion of His Fiftieth Birthday on January 12, 1979*, ed. by E. Saarinen, I. Niniiluoto, and M. B. Provence-Hintikka, D. Reidel, 136–146.
23. "Philosophy and Decision-Making" (first published in this collection).
24. "The Concept of Unreliable Information" (first published in *Studies in Economic Theory and Practice*, ed. by J. Łoś, Państwowe Wydawnictwo Naukowe, now Wydawnictwo Naukowe – PWN, Warszawa – North Holland Publishing Company, Amsterdam 1981, 327–341).
25. "On Defining Information" (first published in *Philosophia Naturalis* 21 (1984), 444–449, now published by Vittorio Klostermann, Frankfurt am M.).
26. "Rationality as a Value" first published in *Dialectics and Humanism* 2 (1984), 289–297, Państwowe Wydawnictwo Naukowe, now Wydawnictwo Naukowe – PWN, Warszawa).

The Editors are also grateful to Anna Szaniawska for permission to reproduce the already published papers and for translating the article "Philosophy and Decision Making". Finally, the editors thank Annie Kuipers, Jaakko Hintikka, Grzegorz Lissowski and Piotr Świstak for their help in preparing the present book.

The Editors express their gratitude to the Institute of Philosophy of the Jagiellonian University, Kraków, Poland, for its financial support in preparing the manuscript of the volume.

Adam Chmielewski *Jan Woleński*

1
SOME REMARKS ON THE PHILOSOPHY OF SCIENCE

During the last few decades, philosophy of science has steadily grown in importance. This is due to many factors, not the least of them being the increasing importance of science itself. Another stimulus has been the belief, shared by many people, that the main business of philosophy is to investigate the sources and validity of knowledge, and that the only genuine knowledge is that provided by science. There is, also, the awareness that philosophy of science is needed, both by the scientist and by the layman who both want to know what are the grounds on which science asserts what it does.

In the present paper I intend to show, with the aid of some examples, what philosophy of science tries to do, to discuss briefly its possible uses, and to define its place in a somewhat broader field, usually called the science of science. To avoid possible misunderstandings, I would like to state that 'philosophy of science' is understood here in a restricted, rather technical, sense, to be distinguished from 'theory of knowledge'. It does not cover, for instance, such fundamental epistemological problem as: what is truth? can it ever be attained? are the sense-data to be trusted as the source of empirical knowledge? etc. Such questions are, no doubt, important and interesting in themselves, but they fall outside the scope of the philosophy of science proper.

To being with, let us recall the well-known semantic distinction concerning the various uses of the word 'science'. First, there is science as a certain type of activity called research. Next, there is science as the product of this activity, contained presumably in books, papers, and reports, but also in some non-semantic material objects, such as hybridized plants, prototypes of machines, etc. Finally, there is science as a social institution, with an organizational structure, hierarchy, budget, administration, and so on. Of these three basic meanings, philosophy of science is concerned mainly with the second (because it analyzes scientific theories) and, to some extent, with the first (because it reconstructs the way theories are justified).

In this respect, philosophy of science seems to be a proper part of a larger complex of investigations, known under the name of metascience, or the science of science. As far as the subject-matter is concerned this is more or less true. But there is a difference in the point of view between the two disciplines which should be noticed.

Historically, the science of science came into being under the pressure of urgent social needs. It had been realized that science is an important factor in shaping the contemporary world – too important, in fact, to be left alone. The gap between what science could do (in social terms) and what it actually did was thought by many people as too great. Motivated by the desire to fill this gap, scientists themselves (cf. the pioneering work of Bernal) began to investigate the potentialities of science. Many practical problems were raised. How does one improve the organization of research? What are the priorities in financing research? How does one measure the effectiveness of scientific work, and by what means can it be increased? The answers to any of these, or related, questions presuppose some knowledge of the actual state of science and of the way it functions. Hence much effort has been devoted to analysing science, its history and the role it plays in society.

Such inquiries quickly developed into a discipline in its own right. But this discipline, the science of science, remained heavily influenced by the initial preoccupation with practical questions that originally made science the object of research. I am far from implying that there is anything objectionable in such a practical approach. If I mention this tendency it is only because I consider it characteristic of the science of science. I also think that the practical approach is what often distinguishes metascience from the philosophy of science. Let us discuss it in terms of an example.

It is commonly agreed that the main difference between scientific and non-scientific activity lies in the methods they each employ. But what are the methods used in science? The last question can be interpreted in two ways. First, one may ask, by what means does the scientist obtain his results? Secondly, how does he justify them? The difference between the two interpretations is not negligible, although they are obviously related. Following Reichenbach's well-known suggestion, we may say that methods of science can be investigated either in the context of discovery or in the context of justification.

Now, the practical concern of the metascientist is mainly with the methods of discovery. He investigates the way new ideas are born because he wants to influence the process. Hence his interest in heuristics, in the psychology of discovery, and also in the history of science. In trying to find out the regularities, i.e. the patterns of discovery, he has a practical objective in mind: he wants to know how scientific output can be increased, both quantitatively and qualitatively. The recruitment and training of research workers, the planning and organization of their work, the distribution of information – all these are considered to be factors which influence the creative process. Some knowledge of the methods applied in this process is a prerequisite for a successful analysis. It is a commonplace to say that the methods in question are elusive and that it is difficult to define them precisely.

The approach of a philosopher of science is somewhat different. His starting point is the product of research, usually in the form of a scientific hypothesis or a theory, rather than the process leading to it. A typical question he asks is, on what grounds can the hypothesis be provisionally accepted? In other words, the philosopher of science deals primarily with methods of justification.

Is there a need to ask such questions? Does not the scientist himself know the answer? Strangely enough, very often he does not. The scientist is usually much too preoccupied with his own work to bother with the somewhat abstract problem of its validity. He is content to follow the 'rules of the game', for he knows they 'worked' in the past (i.e. the results obtained proved reliable), and therefore it is reasonable to assume they will work again in the future. I am, of course, speaking now about the average research worker. Some scientists of the very highest calibre (e.g. Einstein) are known to have given much thought to the problem of the foundation of their theories. This does not change the general picture. Making the methods of justification explicit still remains one of the main tasks of the philosopher of science.

There is a certain danger in this procedure, connected with the delicate question: does the philosopher of science describe what the scientists do, or does he try to tell them what they should do? Neither answer seems quite satisfactory.

The purely descriptive approach in not possible, for reason that have already been hinted at. A scientist does not, typically, validate his assertions according to some explicit and consciously accepted methods. Indeed, such methods hardly exist, in the sense of being clearly stated. This fact by itself does not make the work of the scientist less valuable. There is a close analogy here with deductive reasoning: people have been known to make deductive inferences quite correctly without ever having heard of the logical rules governing deduction and, in fact, before such rules had ever been discovered. It follows, however, that the methodological work of a philosopher of science cannot consist of pure description.

He may, of course, try to abstract from scientific activity discernible patterns and then to say that the majority of scientists behave as if they were applying such and such method of justification. This is easier said than done. Science proceeds in a much less orderly way than is commonly thought, hence the difficulty in ascertaining the above-mentioned patterns. Let us assume, however, that this difficulty is overcome and a hypothetical method has been generalized out of particular cases. A question immediately arises: is this method to be recommended and, if so, on what grounds?

The philosopher of science can hardly avoid asking such question. If he is bold enough to answer it, he is open to the objection that he is attempting to legislate over the scientific activity. For either to recommend or to reject a method is tantamount to saying to the scientist: this is how you *ought* (or *ought not*) to proceed. Philosophers have been known in the past to try to establish that kind of 'control' over science, but it is difficult to see how they could ever justify their claims. This explains the rather distrustful attitude of some scientists towards the philosophy of science.

A way out of this predicament is in what K. Ajdukiewicz has called "the understanding methodology" (*metodologia rozumiejąca*). It can be described as follows:

Any method is a prescription for acting in a certain way; the reason we comply with the prescription is that we hope to achieve something thereby. Hence, the fundamental characteristic of any method is its efficacy with respect to a given objective. The argument for adopting a method is thus relative to the purpose the method is intended to serve, and consists in showing that the method is efficient enough with respect to this purpose, or that it is *the most* efficient in a class of available methods. Efficiency, in turn, may be defined in various ways, depending on the context, but the main ingredient of the concept is the probability of achieving the purpose by the application of the method.

If this is accepted, the adoption of a method in science may, in principle, be justified along these lines. Such a procedure presupposes that the objective is unequivocally defined. Quite often, however, the scientist himself is not clearly aware of the purpose the method is to serve. In such cases a hypothetical justification is usually possible: *if this* is the purpose the scientist has in mind, *then* the method is good, in terms of its efficiency with respect to the purpose.

The virtue of this approach is that it is non-arbitrary. The task of the philosopher of science consist in appraising a method used in science as a means to achieve a hypothetical end. In some cases, at least, this may be done quite convincingly. An example will, perhaps, make the matter clearer.

One of the fundamental procedures in science is that of generalization on the basis of experience, as in the estimation of parameters, the testing of various statistical hypotheses, etc. The procedures in question are usually called inductive (or statistical) inference, and the methods governing them belong to the standard equipment of any scientist. How can their use be justified?

The problem is relatively simple if the scientist has a practical objective in mind (as, e.g., in statistical quality control, when he wants to minimize the loss incurred owing to the erroneous acceptance or rejection of a certain product). The objective is then clearly defined, e.g., in terms of money. And conceivable method makes the decision depend upon the outcome of experiment, such as drawing a random sample and inspecting it. The most efficient method is that which minimizes the expected loss (or maximizes the expected income, which amounts to the same thing). The exact meaning of 'minimization' may be sometimes difficult to define, but once it is defined, it determines the optimum method, and, at the same time, provides the justification for the optimum method, and, at the same time provides the justification for its adoption relative to the given objective. It may be said that the problem of inference has been avoided here, in the sense that no conclusion was arrived at: the only thing that is accomplished is that a choice of action has been made.

Suppose, however, that the research is conducted for a purely theoretical purpose. Which rule of inference is then the optimum one? In order to answer the last question we must provide an exact definition of the purpose, and hypothetically assume that this is what the scientist wants to achieve. Sometimes the reasonable assumption is that he wants to obtain the true conclusion (as in some tests of hypotheses). In other cases his intention may be to minimize the error (point estimation) or he may want to achieve conflicting ends: both to maximize the probability of obtaining a true conclusion and to maximize its exactness (interval estimation). And definition of the scientist's objective consists in specifying his valuation, sometimes called the 'epistemic utility function', of every possible result of his inference. A method of inference may be acceptable relative to one such purpose, and unacceptable relative to some other.

The methodological analysis of inductive inference is actually a lot more complex than it would appear from the above outline. Many questions are still unanswered, and conflicting views on some fundamental issues (e.g. on the use of *a priori* probabilities) have been formulated. There is no need to go into such technical details here. The idea was to give an example of how present-day philosophy of science approaches the problem of scientific methods. A characteristics feature of this approach is its application of the theory of decision-making. This last theory provides the conceptual framework within which scientific methods (some of them, at least) can be appraised as means toward a given end, practical or theoretical one.

The tools provided by the theory of decision-making made it possible to analyze scientific methods in an exact way: statements like "this method is (or is not) good" have achieved a clear meaning, becoming provable thereby. This is an important advantage.

As a by-product of our analysis, we gain insight into some difficult meta-scientific problems, e.g., the distinction between pure science and applied science. On the view advanced here, the difference consists mainly in valuations.

Valuations do play an important rôle in all scientific work, although this has not always been openly admitted. Philosophy of science based on decision theory explicitly introduces valuation into the description of scientific work. Now, in applied science the valuations influencing the choice of a particular method are determined, naturally enough, by the practical goal of research. In the case of pure science it is less easy to determine the valuations that govern the research. The notion of theoretical, or purely cognitive, purpose of research is somewhat vague. It can be made more precise, but any attempt to do so is, within certain limits, arbitrary. Thus, for instance, it is customary to equate the theoretical 'loss' incurred in estimation with the square of the error. But there is nothing inherent in the procedure that compels us to do so: other strictly increasing functions of the size of error would perhaps do equally well.

Some philosophers of science maintain that this proves theoretical superiority of applied over pure research: there is less arbitrariness in procedures directed towards a practical goal. I mention this view because of its paradoxical character. For a very long time pure science was regarded as something much more dignified than applied science; the relation seems now to have been reversed. The relative merits of the two types of research can not, of course, be reduced to the above standard of comparison. It is, however, interesting to note that the traditional view has been undermined by the analysis of scientific methods in the light of decision theory. Thus philosophy of science has contributed to the much-debated issue of whether the results of pure science are more firmly established than those of applied science.

Another interesting methodological problem that has been raised in this context is the following one. There is a conflict between the immediate and the long-range efficiency of scientific methods. The decision theory has shown that in applied research the method connecting experimental findings directly with a choice of action is more efficient, relative to the purpose of research, than the method which consists in drawing a conclusion from the data and basing the choice of action on this conclusion. On the other hand, if the conclusion *is* drawn, it can be used repeatedly in practical problems that may arise later on. It is true that the applications would be less efficient than the procedure of going directly from data to the practical decision; but their very number would compensate for the loss in efficiency.

Hence, it very often pays to gather knowledge for no immediate practical purpose and 'store it away' for further use, not specified in advance. In which cases is it reasonable, in practical terms, to follow this course rather than make immediate use of data? There is no conclusive answer to this important question.

Scientific methods are not the only concern of the philosophy of science. Another group of problems are those arising in connection with the meaning of scientific statements and the structure of theories. These two topics are closely connected, for obviously the meaning of scientific statements is determined, to a large extent, by the role they play in theories.

Again, it must be realized that such problems are seldom investigated by scientists themselves. Typically, a scientist develops a theory without giving much thought to its logical status. This explains the puzzling fact that the exact meaning of statements of a theory is seldom made precise in the exposition of that theory.

Thus, for instance, it is quite often unclear whether a statement ought to be considered as an implicit definition of the terms it contains, or an assertion making use of those terms on the assumption that they obtain their meaning from some other source. The difference is highly important, for in the first case the statement is tautological, hence incapable of being falsified; in the second one, it is factual and subject to empirical test.

Such questions are very difficult to answer because of the great complexity of most scientific theories. What the philosopher of science tries to do is to 'reconstruct' the theory, which means, roughly: to formulate the logical relations between its constituent statements and establish a partial ordering of them, from the highest-level hypotheses to direct empirical findings. The tools needed for such a task are provided by modern logic, particularly by what is called the theory of models.

Because of the above difficulties, this is more a program than na achievement. Only the simplest theories have been successfully analysed in such way. The results show that the distinction between analytic and synthetic statements in science is more elusive than is commonly thought. There are reasons to suppose that no part of a scientific theory is purely tautological; which makes it even more difficult to differentiate between what is 'true in virtue of a terminological convention' and what is true in the light of factual data.

Another problem which belongs to this group is how the theoretical concepts of science, such as 'electric-field vector' or 'wave function', obtain their empirical meaning. Until recently it had been thought that they could, in principle, be defined in terms of directly observable entities. The view now prevails that they cannot be so interpreted. The main reason is that if such were the case, the theory would not be open to further extensions. This has been shown on some artificially simple examples. It remains to be seen that the assertion applies to actual scientific theories. If it does apply, the empirical meaning of scientific concepts would be determined by the fact that the statements in which they occur have empirical consequences.

Other problems arise in those disciplines which derive their terminology from everyday language. Here semantic analysis is perhaps most needed; because of the notorious vagueness of natural languages, the clarification of concepts becomes a necessary prerequisite of any significant progress.

Social sciences, for instance, make free use of concepts denoting dispositions of some kind, like 'intelligence', 'radical prejudice', etc. Attempts to make these concepts more precise have run into logical difficulties of a very general character. The co-operation of social scientists and philosophers of science has resulted in various formal models connecting a disposition with its symptoms (i.e., a set of observable properties). Such models are now being further developed. The testability of assertions made in the social sciences depends upon the success of these attempts.

To recapitulate: the main concern of the philosophy of science is with scientific methods and the structure and meaning of scientific theories. Philosophy of science tries to make explicit the methods of research and justify their use in terms of the purpose the scientist wants to achieve. It also investigates the formal aspects of scientific theories, paying particular attention to their conceptual apparatus. On this view the philosopher of science makes extensive use of modern logic and related disciplines like theory of decision-making.

The problems that philosophy of science deals with are no doubt interesting in their own right. One may ask, however, what profit a scientist can expect from a logical analysis of the product of his activity. First, he can gain a better understanding of what he is actually doing, and this would help him in his work. Also, the philosophy of science promises to counter-act the undesirable effects of narrow specialization. If anything integrates present-day science, it is its method. Philosophy of science brings this fact to the attention of the scientist, enabling him thereby to put his work in a larger perspective.

2

INFORMATION AND DECISION-MAKING AS TOOLS OF PHILOSOPHY OF SCIENCE*

I INTRODUCTION

This is an essay in pragmatic methodology. The distinction between pragmatic and apragmatic methodology seems to be, by now, well established. Broadly speaking, apragmatic methodology treats science as a system of statements and investigates its properties: structural relations between statements, semantic characteristics of the language, meaning postulates for its theoretical concepts, etc. The analysis is, in a sense, 'impersonal', i.e. it disregards the fact that science is the product of scientist's activity.

Such an approach is, of course, perfectly legitimate and helpful in discovering important features of scientific theories. There is however, another group of problems arising when the role of the scientist is taken into account. Certain new terms are thus added to the methodological vocabulary, such as: acceptance of statement, its (subjective) probability, the goals directing scientific research, etc. They are distinctive of pragmatic methodology.

Decisions is obviously a pragmatic concept. It presupposes an evaluation of the different courses of action which are to be decided upon. In the case of science, the valuations reflect the goals of scientific activity. One of these goals is to acquire information.

The above remarks explain the purpose of the present paper. It is an attempt to show in broad outline how the concepts of decision and information can be employed in the pragmatic methodology of science (and are, in fact, being employed; see [5] as a classical work of this type). In order to do that, we shall have to specify the meanings, of these crucial terms. This, in turn, will demand the introduction of some formalism. Technicalities however, will be kept down to the necessary minimum, even at the cost of simplifying the account. It is assumed that the reader is more interested in the general idea itself than in its purely formal aspect.

II QUESTIONS

One of the basic notions we shall need is that of a question. Out of the large variety of questions which are being asked in science, we shall consider only those for which the set of proper answers is well defined. "Proper" means: direct and precise. In other words, we are concerned with question of the type: "which s?", where the set of all possible s is known in advance. This undoubtedly imposes a severe limitation on the scope of our inquiry. We shall come back to this problem when discussing the applicability of the model.

Formally, a question can be represented by the set of its proper answers. We will denote it by S, thus assuming that the elements of S are jointly exhaustive and pairwise exclusive (because exactly one answer to the question is true). It is assumed, for simplicity, that S, and all other sets that will be later introduced, are finite. Now, it rather seldom happens that the scientist is in a position to obtain a direct (and true) answer to the question S. In most cases, he will obtain knowledge about S indirectly, through some kind of evidence, belonging to another set, say X, of pairwise exclusive and collectively exhaustive elements. Typically, X is the set of possible outcomes of an experiment, so designed that it will have some bearing on the original problem S.

Obviously, of crucial importance is the relation between the elements of X and those of S. It will, in general, be probabilistic and represented by the joint probability distribution function p on $X \times S$. The function must satisfy:

$$p(x, s) \geq 0 \quad \text{for } x \in X, s \in S; \qquad \textstyle\sum_x \sum_s p(x, s) = 1. \tag{1}$$

For simplicity's sake, we leave here out of account the weaker type of relation between X and S (cf. Szaniawski [7]). It can be described by the assertion that each s determines a probability distribution, say $p(x \mid s)$, on X. We thus would have

$$p(x \mid s) \geq 0 \quad \text{for } x \in X, s \in S; \qquad \textstyle\sum_x p(x \mid s) = 1 \quad \text{for } s \in S. \tag{2}$$

Relation (1) is stronger than (2), because it generates (2) through the definition

$$p(x \mid s) = p(x, s)/p(s), \text{ where } p(s) = \textstyle\sum_x p(x, s). \tag{3}$$

On the other hand, (2) is considered by some people to be more realistic than (1); it is maintained that the marginal probability distribution $p(s)$ on S (usually called *a priori* distribution) is, in most applications, unknown or even meaningless. And it is this *a priori* distribution which would make (2) equivalent to (1). The discussion of relative merits and demerits of the two approaches would leads us too far away from the main topic. Anyway, the stronger assumption makes possible a more unified treatment. This is the main reason for our choosing this approach (called Bayesian) for the present exposition. Let us, therefore, consider for a given S the triple

$$\langle X, S, p \rangle \in Q(S), \tag{4}$$

where p satisfies (1) and $Q(S)$ is interpreted as the set of all $\langle X, S, p \rangle$ with S fixed. By specifying p accordingly, we obtain two important special cases.

First, let us impose on p the additional condition:

$$\text{For each } s, \text{ there is exactly one such } x \text{ that } p(x, s) \neq 0, \tag{5}$$

or equivalently

For each s, there is such x that $p(x \mid s) = 1$ (5′)

If p satisfies (5) then X generates through p a partition $\{S_x\}$ of S, where for any x, S_x is the set of all such s for which x has the conditional probability equal to 1. It follows that if x is obtained then the true s belongs so S_x. The knowledge concerning S, due to X, is no longer probabilistic: X can be interpreted as the question, which S_x does the true s belong to?

Suppose now that p satisfies the following condition.

For each x, there is exactly one such s that $p(x, s) \neq 0$, (6)

or equivalently

For each x, there is such s that $p(s \mid x) = 1$. (6′)

Under this condition, each x specifies s uniquely. We thus obtain the original question, "which s?", as a special case of (4), through the condition (6) imposed on p.

In view of the above considerations, we will generalize the concept of question, reading $Q(S)$ in (4) as: question concerning S. An element of $Q(S)$ satisfying (5) will be called 'categorical question concerning S', while (6) defines 'perfect question concerning S'. The choice of terms is self-explanatory.

III INFORMATION

"The basic idea that underlies the concept of information (. . .) remains essentially the same in all cases. The information of s is the amount of uncertainty we are relieved of when we come to know that s is true". This statement by Hintikka ([4], p. 312) seems to be an accurate account of the idea behind the different concepts of information. It follows that information contained in a statement ought to be a decreasing function of its probability.

Two such measures of information are well known:

$$\mathrm{inf}(s) = -\log p(s), \tag{7}$$

$$\mathrm{cont}(s) = 1 - p(s). \tag{8}$$

The first one is mostly used in communication theory, the second one plays an important role in the theory of semantic information, as developed by Carnap and Bar-Hillel some twenty years ago (cf. [1] and [2]). Each of the two concepts of information has certain desirable properties which justify its adoption. They are discussed in Hintikka [4]. The same author suggests certain derived measures which make it possible to define or at least justify such well known methodological concepts as 'evidential support' or 'explanatory power'. They are discussed, ultimately, in terms of either inf or cont.

It is rather natural to evaluate questions, as defined by (4), in terms of information: how much information do we gain by having the question answered, i.e. by getting to know which element of X is true? Equivalently, by how much is the uncertainty concerning the true element of S reduced, due to the answer to some given $Q(S)$? If information is interpreted as inf, we obtain the Shanon formula

$$I(X, S, p) = \sum_s p(s)\text{inf}(s) - \sum_x p(x) \sum_s p(s \mid x)\text{inf}(s \mid x), \tag{9}$$

measuring the average reduction of uncertainty in terms of inf. It is equal to the difference between the so-called entropy of S and the average, over all x in X, of the conditional entropies of S, given x.

Analogous formula, say T, for cont is

$$T(X, S, p) = \sum_s p(s)\text{cont}(s) - \sum_x p(x) \sum_s p(s \mid x)\text{cont}(s \mid x) \tag{10}$$

and it behaves, relative to the changes in p, in a very similar way to I.

The role of I or T is that they make possible to compare questions concerning the true s in S. Obviously, there are better questions and worse questions in $Q(S)$; formulae (9) and (10) give exact meaning to such comparisons. Intuitively, the value of I and T ought to be minimal in the case of independence between X and S, and maximal when the question is perfect. And, indeed, this is the case. Needless to say, we are not always in the position to have perfect questions answered.

IV DECISION

All the concepts so far introduced were apragmatic, unless p was interpreted as subjective probability, i.e. the estimate, by the subject, of chances of the respective statements to be true. The last remark should be emphasized. It is an increasingly popular belief that this is *the* way to interpret probability, science being no exception. One argument for the subjective interpretation is that it makes probability applicable to any kind of statements (unlike the frequency concept) and independent of the language (unlike the logical probability). If this view were adopted, all methodological concepts defined in terms of probability would become pragmatic.

There is no need here to take a position on this much debated issue. Whatever the status of information (as defined above), the concept of decision is clearly pragmatic, if only because it presupposes a preference ordering of the alternatives.

The paradigm of decision problem is based on the assumption that there is some uncertainty as to the consequences of decision. Alternatively, that the subject does not know all he needs to know about the conditions in which his decision takes place. To be sure, this assumption can be dispensed with, and indeed there are situations in which the decision maker has all relevant data. Then, however, his problem becomes trivial, except for possible computational difficulties in finding the optimal alternative.

The above remarks presuppose that the subject's valuations of all the possibilities are given, on a one-dimensional scale of preference. Another assumption is that if he does not know the relevant conditions (the actual 'state of the world') he at least can enumerate the possible states of the world. Finally, if he knows the alternatives, call them 'actions', he is ready to make a choice.

There is no doubt that these are pretty strong demands on the description of decision problem. It is difficult to see how they could be weakened, without making the problem of choice logically unmanageable.

The above remarks are summarized by the statement that decision problem, call it U, is defined as

$$U = \langle A, S, u \rangle, \tag{11}$$

where A is the set of actions, S is the set of possible states of the world, and u is a function (called 'utility'), representing valuation by the subject of all possible pairs (a, s), where $a \in A$, $s \in S$.

It is not accidental that the symbol S denotes the set of possible states of the world (or, rather, their linguistic counterparts). Here the same assumption is made as in the case of questions: exactly one state of the world obtains, i.e. exactly one of the elements of S is true.

Having a question concerning S answered will change the decision problem; in general, it will improve the decision by removing some uncertainty concerning its consequences. To put it the other way round, a question $Q(S)$, may be evaluated in terms of U, provided that it is relevant to U. The relevance consists precisely in the fact that the same S appears in (4) and in (11).

The principle of evaluating a question with respect to decision problem is easy to grasp. It has been explicitly discussed elsewhere (see e.g. [6] or [8]). The account below will, therefore, be brief.

The obvious idea is to calculate the (average) increase of utility, due to having the question answered, relative to the 'status quo' level of utility, i.e. to what is obtained when the decision is made without such support. This is done as follows.

If no additional information is forthcoming, the best that the decision maker can do is to choose such a in A which maximizes the expected value of u, under the probability distribution $p(s)$ on S. This gives the utility

$$\text{Max}_a \textstyle\sum_s u(a, s)p(s). \tag{12}$$

Suppose now that the subject, before making his decision, will have a question of the type $Q(S)$ answered. It will then be possible for him to make his choice of a dependent on the answer received, i.e. on x. Graphically, the situation can be represented by the diagram

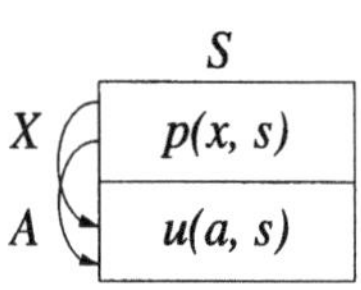

where the upper rectangle represents the matrix of joint probabilities, the lower rectangle represents the matrix of utilities, and the arrows indicate a function from X to A, interpreted as: if x is obtained then a will be chosen. Of course, from each row in the upper matrix there will be an arrow towards a row in the lower one; the drawing is incomplete, for the sake of clearness.

Also, there is a considerable number of such functions, each of them representing a possible way of using the fact that the question has been answered. The best way of utilizing answers to the question is, of course, to associate with each x such a for which the expected utility, relative to $p(s|x)$, is maximal. Averaged over all x in X, this will give:

$$\sum_x p(x) \operatorname{Max}_a \sum_s u(a, s)p(s \mid x). \tag{13}$$

The value, with respect to U, of the question concerning S is then obtained by subtracting (12) from (13). The result, call it C, is the expression

$$C(X, S, p; U) = \sum_x \operatorname{Max}_a w(a, x) - \operatorname{Max}_a \sum_x w(a, x), \tag{14}$$

where $w(a, x)$ is a convenient shortening

$$w(a, x) = \sum_s u(a, x)p(x, s). \tag{15}$$

The expression C, as defined in (14), represents the improvement of decision, due to having the answer to the question $\langle X, S, p\rangle$. In another terminology (see [8], it has been called 'pragmatic information' (contained in the question of the type $Q(S)$).

V INFORMATION AND DECISION IN SCIENCE

The concepts sketched above are sufficiently general to cover all types of decision-making. What is the distinguishing feature of decisions made in science?

It might be said that the decision-theoretic framework is particularly well-suited to applied science. If the immediate goal of research is to improve a particular decision about to be made, then all pieces of this conceptual apparatus find a natural interpretation. Elements of A represents contemplated courses of action, monetary gains and losses play a decisive role in determining the utility function u, and the scientist's task is to provide an answer to a question $Q(S)$ which, in turn, can be evaluated in terms of the decision problem.

Our primary concern, however, is with science as purely cognitive activity. Science, so conceived, results in the acceptance of certain statements. It follows that, in this case, the basic type of decision is: which statement to accept?

This determines, in broad outline, the character of the set A: its elements are truth-functional compounds of the elements of S. The simplest possibility is to interpret them as disjunctions of some s in S. The idea is due to Levi (cf. [5]) who also points out that if the decision is to accept a disjunction, say, $s_1 \vee s_2$ it would be inconsistent not to accept the disjunction $s_1 \vee s_2 \vee s_3$ which is implied by the former one. Hence, the problem really consists in deciding which disjunction to accept as the strongest.

Speaking more formally, A is a family of subsets of S,

$$A \subseteq 2^s \tag{16}$$

satisfying certain conditions:

(i) all elements of A are non-empty;
(ii) A has at least two elements, such that none of them contains the other (otherwise there would be no choice);
(iii) each s belongs to at least one a (otherwise, for some s, no a would be true).

Three cases are of special importance:

(a) $A = S$, i.e. only the strongest conclusions are admissible.
(b) $A = 2^s - \{\varnothing\}$, i.e. all conclusions are admissible.
(c) A is a partition of S, i.e. any two conclusions are incompatible (case (a) may be considered an instance of (c): every element of the partition contains just one s).

Case (b) has been investigated by Levi in [5], under certain assumptions concerning the utility function. Cases (a) and (c) are simplified models of certain procedures in mathematical statistics: (a) corresponds to point estimation, (c) to the testing of hypotheses (somewhat generalized).

It is up to the scientist to determine the exact nature of his set of admissible conclusions. He is guided in this by the needs of his research. Thus, for instance, he may be interested in finding out whether a particular s in S is true or not. His A then consists of two elements: $\{s\}$ and S-$\{s\}$. Conclusions of this type are well known from statistical induction.

What about valuations, characteristic of pure research? They are usually called 'epistemic utility', following a suggestion by Hempel in [3]. And they reflect the goal of research. The plural is not accidental. In Hintikka's words, 'in general, the scientific search for truth is much less of a single-goal enterprise than philosophers usually realize' ([4], p. 321). The objectives differ, sometimes they are completely opposed and must be reconciled by some kind of compromise. In other words, given S and a set A of admissible conclusions based on S, there can be more than epistemic utility function U, defined on $A \times S$. We will give a few examples.

Probably the simplest aim of research is just to accept the true conclusion. All errors are then considered of equal importance and the same goes for all cases of being right. The utility is, therefore, either zero or one, and if A is defined as in (a) we get

$$u(a, s) = \begin{cases} 1 & \text{if } a = s \\ 0 & \text{if } a \neq s. \end{cases} \tag{17}$$

With respect to such an epistemic problem (call it E), the value of a question on S, as defined by (14), is

$$C(X, S, p; E) = \textstyle\sum_x \mathrm{Max}_s\, p(x, s) - \mathrm{Max}_s \sum_x p(x, s). \tag{18}$$

The expression (18) has properties remarkably similar to (9) or (10); this is shown in more detail in [8]. Which means that evaluation of a question in terms of information, or in terms of the epistemic problem (17), yields essentially the same results.

It ought to be noted that if the same epistemic utility is applied, *mutatis mutandis*, to a problem of type (b) the optimal conclusion in A is the disjunction of *all* elements of S, tantamount to no conclusion at all (because the truth of the maximal disjunction had been presupposed). And S remains optimal, no matter what evidence is provided by a question $Q(S)$. Hence, the value C of any such question is zero, relative to an epistemic problem of this type.

The above consequence ceases to be puzzling when we realize that if truth is the only thing we want, then the possible conclusions are appraised in terms of their probability, and the optimal conclusion is the 'safest' one, i.e. tautological with respect to the knowledge at hand. Provided, of course, that such a conclusion belongs to A; and, under (b), it does. Any research concerning S thus becomes valueless.

If we want to avoid such a nihilistic view, and still have A defined as in (b), we must change the utility function, introducing some other factor than the search for truth, e.g. Levi's "relief from agnosticism". This is the basic idea behind his "gambling with truth".

At the same time, it provides another example of epistemic utility. According to Levi, u is defined as follows.

$$u(a, s) = \begin{cases} -q \cdot m(a) & \text{if } s \in a \text{, i.e. when } a \text{ is true} \\ -q \cdot m(a) & \text{if } s \notin a \text{, i.e. when } a \text{ is false,} \end{cases} \tag{19}$$

where $m(a)$ is the relative number of elements in a, and the coefficient q, satisfying $0 \leq q \leq 1$, is the weight attached by the scientist to the strength of conclusion, as measured by $-m$. Acceptance of a conclusion is thus a compromise between two conflicting aims. One of them is to avoid error; the other is to obtain an informative statement (the measure of informativeness being provided by the negative function m). Obviously, this u differentiates both errors and non-errors, as to their importance; and every error has lower utility than any non-error, which expresses the precedence of truth before informativeness in the scientist's hierarchy of values.

Parametric utility (19), when applied to a decision problem of type (a), reduces to (17): if $m(a) = 1/n$ for all a, then all errors (and non-errors) are of equal importance.

A natural modification of (19) would consists in replacing the uniform and regular measure m by probability p;

$$u(a, s) = \begin{cases} 1 - q \cdot p(a) & \text{if } s \in a \\ - q \cdot p(a) & \text{if } s \notin a, \end{cases} \tag{20}$$

where $p(a) = \Sigma_s\, p(s)$. Informativeness of a would then cease to be independent of adopted probability measure; it would become a decreasing function of p

(in fact, simply minus p), in accordance with the general idea of information (cf. above).

The meaning of (20) becomes clearer when we let the parameter to q assume its extremal values. For $q = 0$, (20) reduces to (17). The consequence is not surprising: $q = 0$ means that informativeness of a is of no account, therefore the only remaining goal is a's truth.

On the other hand, if $q = 1$ (20) can be rewritten as follows:

$$u(a, s) = \begin{cases} \text{cont}(a) & \text{if } s \in a \\ \text{cont}(a) - 1 & \text{if } s \notin a \end{cases} \tag{21}$$

which means that both true and false conclusions are ordered preferentially according to their content, under the condition that any truth is better than any falsehood.

This is reminiscent of Hintikka's idea to "maximize expected content" (cf. [4], pp. 325–328), and indeed the epistemic utility defined in (20) leads to a generalization of Hintikka's results. The measure of content of conclusion, i.e. a variant of information, turns out to be a measure of the degree to which the goal of research has been achieved. That is, if what Hintikka calls "global theorizing" guides the scientist's efforts. Another possibility is "local theorizing", which would require a different description in terms of u.

Finally, it ought to be remarked that truth and information are not the only criteria for appraising results obtained in science, even if science is conceived as purely cognitive activity. Other known candidates are: simplicity, originality, depth, aesthetic qualities ('elegance'), etc. There is place for them within the general framework of epistemic utility. So far, however, no attempt to explicate the notions with sufficient precision has been successful.

These examples will suffice as an illustration of the possibilities offered by the concept of epistemic utility. Their formal implications will not be further developed in this paper.

VI A COMMENT

What is the significance of the above concepts for the philosophy of science? A partial answer has already been sketched.

They are an attempt to explicate certain notoriously vague notions, basic for understanding of science as goal-directed activity. First, the goals are specified in various ways, then the scientist's decisions are accounted for, under the assumption that these are, in fact, the objectives of his research.

The decisions concern, primarily, the choice of a statement from a set of admissible conclusions. Indirectly, however, they concern the interpretation of possible evidence in X: from the class of all rules of inference from X to A (Abraham Wald would call them decision functions), the scientist selects the one which is optimal in the sense that it maximizes his epistemic utility. The rule of inference is thereby justified as an efficient way of achieving the purpose of research. Also, different types of evidence (different questions from the class

$Q(S)$ on S) can be compared in terms of the increase of u, under the assumption that they are optimally used.

It should be emphasized that all results of the kind described above are obtained relative to a simple model, with fairly strong assumptions. Direct applicability of the model to scientific practice raises serious doubts. It is only rarely the case that all the sets in question are well-defined and known, just as all the numerical data which are needed, the joint probability distribution of X and S in the first place.

The objection is justified. However, as Levi ([5], p. 46) rightly says, "it should be noticed that, in a systematic account of conditions of rational belief, not all principles involved are directly applicable in real-life situations. A principle whose scope of applicability is extremely limited might still be of central importance for deriving and otherwise systematizing criteria that are more directly applicable."

Philosophy of science has always proceeded on this assumption. Thus, e.g., investigation of conceptual structure of scientific theories in terms of model theory has as its premiss a highly simplified description of the empirical system's language. As a consequence it turns out that, although the investigation yields a number of results concerning, say, the analyticity of statements, it is hardly ever possible to determine the analytical component in an empirical theory actually existing. Nonetheless it is usually felt that we gain some insight into such theories by the analysis of their simplified analogues. If such a procedure is useful in apragmatic methodology, there is no reason why it should not be useful in pragmatic one.

Admittedly more serious objection says that significant results in science are not arrived at in the way described above. In this model, all the concepts are defined in terms of a closed set of alternatives. However, most problems in science are open. Progress in science is not so much a matter of choice but of invention. At least part of science consists in creative activity, in which even conceivable results are not known in advance.

All this is quite true, and it would be naive to expect that any kind of algorithm, decision-theoretic or otherwise, will account for scientific discovery. However, scientific research is a complex activity and to a large extent it consists in solving problems of choice. The expression '*experimentum crucis*' signals the need to choose between a hypothesis and its negation as it has been realized quite a long time ago. More generally speaking, inductive inference can be profitably analyzed in terms of decision making. And there is little doubt that the development of inductive methods is one of the signs of scientific progress. Insofar as it helps to understand this part of scientific activity, the decision theoretic approach has a legitimate place in the philosophy of science.

NOTES

* The paper was originally read to a symposium, held by the Institutes of Philosophy of the Universities of Copenhagen and Warsaw at 'Refugiet Fuglsang', Denmark, on the 30th and 31st of March, 1974.

REFERENCES

[1] Y. Bar-Hillel and R. Carnap, "Semantic Information". *Brit. J. Phil. Sci.* 4(1953).

[2] R. Carnap and Y. Bar-Hillel, "An Outline of a Theory of Semantic Information." *Tech. Report* No 24, Research Lab. of Electronics, MIT, 1952.

[3] C.G. Hempel, "Inductive Inconsistencies", *Synthese* 12(1960).

[4] J. Hintikka, "The Varieties of Information and Scientific Explanation", in: *Logic, Methodology and Philosophy of Sciences III*, van Rootselaar and Staal ed., North Holland, 1968.

[5] I. Levi: *Gambling with Truth. An Essay on Induction and the Aims of Science*, Alfred A. Knopf, 1967.

[6] K. Szaniawski, "The Value of Perfect Information", *Synthese* 17(1967). [see this volume, pp. 140–153.]

[7] K. Szaniawski, "Questions and Their Pragmatic Value", in: *Logic, Language and Probability*. R. J. Bogdan and I. Niiniluoto ed., D. Reidel, 1973. [see this volume, pp. 154–156.]

[8] K. Szaniawski, "Two Concepts of Information", *Theory and Decision* 4(1974). [see this volume, pp. 157–166.]

3

METHOD AND CREATIVITY IN SCIENCE

The apparent conflict between standardization in research, due to the use of various methods as established procedures, and the uniqueness of a scientific discovery, are discussed in the article. It is indicated that this paradox may be eliminated if the differentiation of methods as to the degree of definiteness is taken into account. The role of methods at the various levels of goals which research is to attain is analysed. From that point of view the division of labour, as seen in present-day science, is of special significance. Intertwining of inventiveness and method-based procedures proves a characteristic of scientific research.

The concepts of science and method are so closely intertwined that links between them are not questioned by anyone. It is even believed that if there is something that singles out science among other spheres of human mental activity it is the nature of the methods used in it. It is true that methods do change with time. That they are being improved, and sometimes discarded, and that new elements are more or less regularly being added to the set of accepted methods, yet at a given time it is the methods which decide whether a given statement makes part of the body of science. The task of the methodology of sciences is to codify those methods and to study the role they play in man's cognitive activity.

The above opinions are widely held, and yet the formulation 'the method of pursuing science' has a ring of a paradox. That paradox is attributable to the fact that the pursuit of science is a *sui generis* creative activity, and creative activity by definition cannot be subordinated to methods. Expressions like 'the methods of writing lyric poetry' or 'the methods of composing symphonies' seem grotesque. A method of doing something determines that which is to be done, whereas the very nature of creative activity consists in its being unrestricted and in the freedom of making choices. A method imposes a certain pattern of action, whereas the more a creative activity deviates from a pattern, the more it is valued. Creative activity yields unique results, whereas the use of methods leads to a replication of patterns.

The conflict between the standard-setting function of methods and the freedom of invention seems to account for the objections raised by scientists when it comes to the methodology of sciences. They claim that it is at best superfluous, since, by its very nature, it cannot answer the question which is the most important to the scientist, namely: how to make discoveries? Instead it suggests the codification of those operations which a scientist knows how to perform (such as inference, definition, measurement, etc.).

These objections deserve attention, since even if they are due to verbal misunderstandings, they are nevertheless too common to be dismissed as irrelevant. It would, of course, be naive to think that the distortion of communication between the scientist and the methodologist is due to poor semantics only. But when we strive to eliminate communication problems, no harm will be done if we realize those semantic distinctions which are rather easily forgotten.

Let us begin with the term *method*. Praxiology, a discipline in which this is one of the basic categories, defines 'method' as 'a procedure that can be used systematically, that is, in all cases of a given type'. And 'procedure' means more or less 'the choice and order of component operations that combine to form a complex operation'.

The above definition does not determine the degree of definiteness of such a procedure. On this interpretation, both directions which leave no place for freedom of choice, and general instructions which a given person is to follow in a given case according to circumstances and his own preferences would all be called methods. By interpreting the concept of method very broadly, praxiology follows the linguistic usage according to which the term *method* is used with reference to rules which differ widely from one another not only in content but also in form.

If we consider the degree of definiteness, the special form of a method is to be found in algorithm. That term, which was originally used only with reference to mathematical procedures, can be extended so as to cover all such procedures which are fully defined. Using an algorithm thus consists in mechanically carrying out instructions and does not require comprehension of the meaning of the operations performed (i.e., the relation they bear to the goal to be attained). Hence an algorithm can be carried out by a computer.

The formal structure of an algorithm is usually fairly intricate, precisely because of the full definiteness of such method. All possible choices to be made in the successive steps of the whole operation must be foreseen, and a decision (criterion of choice) must be given for each case. And since the choice of an operation in the n-th step depends on the sequence of the previous choices, carrying out of an algorithm implies following a strongly ramified tree (in the sense of a graph), and the concept of algorithm turns out to be, from the formal point of view, identical with that of a strategy (in games with perfect information).

From that point of view with which we are concerned here, the concept of algorithm is important not so much because of the frequency of algorithmic procedures in sciences (even though that frequency is quite considerable, which is shown by the growing application of computers in research), but because of its standard nature. As a completely defined method, algorithm represents a certain ideal type which methods which are most often used in science are approaching in varying degree.

It must be emphasized that the distance between methods used in science and the standard of complete definiteness happens to be considerable, which leaves the researcher with a scope for free decisions. Even almost algorithmic rules as the methods of statistical inference are in fact classes of completely defined methods from which the researcher makes his choice. For instance, in the verification of hypotheses such decisions as the choice of the level of significance, which affects the acceptance or rejection of the hypothesis involved, and the size of the experimental basis of that procedure, i.e. the size of the sample and the way of sampling, must be made by the researcher.

There also are instructions formulated in such general terms as the hypothetic-deductive method, which consists in formulating hypotheses and deducing immediately verifiable consequences from them. In this case very much is being left to the ingenuity of the scientist who uses that method.

These comments do not, however, dispel completely the doubts raised at the outset. For if we value inventiveness in science – as we do in the case of all kinds of creative activity – then the fact that inventiveness in science is only restricted to some extent merely mitigates the objection raised against the 'methodological' interpretation of science, but does not refute it completely.

How are the methods used in science justified? Such a justification usually assumes the form of the proof that, from the point of view of the goals which sciences sets itself, the use of the procedures indicated by the methodology of sciences is effective, and hence rational with respect to those goals. This in turn requires a distinction between the goals which are attainable by method-based operations and those which are not.

It must be recalled here that the formulation of a programme of research intended to justify scientific methods in such a way is due to Kazimierz Ajdukiewicz (1890–1963). The task which he set for the methodology of science was twofold.

First, we have to reconstruct those methods which are actually used in scientific activity, for it is usually not the case that scientists consciously follow certain rules which could be formulated in an explicit manner. The assimilation by a young scientist of those procedures which are specific to a given discipline rather takes place through imitation, which, if necessary, is corrected by his teacher, that is, more or less in the same way as we learn our first language or assimilate the manners prevailing in our milieu.[1] This statement does not in the least discredit the scientists: nor does the statement that Newton did not (because he could not) know the rules of formal logic discredit his deductive reasoning. Hence the reconstruction of research methods through an analysis of the procedures used by scientists in practice is the first task of methodology. That task is neither the only, nor an easy one.

The second point, which is essential in Ajdukiewicz's programme, is to demonstrate the rationality of such methods. The rationality of a method means its effectiveness with respect to a specified goal. Hence at this point the task of the methodologist is to ascribe to a given scientist a specified goal (of which the scientist need not be fully aware) and to prove that a given method is effective with respect to that goal – in the sense of guaranteeing the attainment of that goal, or at least making its attainment highly probable.

The idea outlined above, the advantages of which as compared with earlier approaches will not be discussed here, sets the methodology of sciences a reasonable and attractive task. But here we face the problem: are the goals which we ascribe to scientific activity such that effective methods of attaining them can be formulated?

After all, it is said – in most general terms – that the task of science, on the one hand, is to pose problems and to solve them (a solution consisting in demonstrating that it *is* an answer to the problem under consideration in a given case). Posing and solving problems both require inventiveness, which raises difficulties mentioned at the beginning of this paper. On the other hand, we can easily point to the large variety of methods which are actually being used in research and which require justification in the sense outlined by Ajdukiewicz.

The resolution of this conflict is to be found in the complexity of science. Science is marked by a hierarchy of goals which are subordinated to the principal task as formulated above. Partial goals recur in various contexts and reveal sufficient formal similarities to enable us to subsume the operations by which they can be attained under a single method. For instance, determination of the chemical composition of a mixture is required when we have to solve various problems (not only chemical ones); interviewing selected persons is part of many sociological and psychological studies, etc. Hence we speak about the chromatography method or the interviewing method when we have in mind a certain repeatable procedure.

We can find here certain analogy with digital computers: computation of, say, the determinant is a subroutine which is included in various programmes according to which the computer works. But we have to note an essential difference. In case of the computer, the programme which includes such a subroutine is also an algorithm, so that the computer operates only on the principle of the complete definiteness of a given method. In research it is different. Even if some research procedures follow a scheme which brings them close to an algorithm, or even if it is an algorithm, the way in which the results obtained in that way are used need not fall under any scheme.

The formulation 'need not' is deliberately cautious, for it also happens that in a given case all research simply consists in the application of a set of methods, which obviously requires competence, but does not tax the researcher's ingenuity. Does this discredit such research as non-scientific? Not necessarily so.

Science is a collective undertaking, with its own division of labour (which is spontaneous rather than controlled), and some results obtained by a person X following certain routine operations (i.e., such which depend on a set of methods), may turn out to be necessary for an ingenious hypothesis formulated by a person *Y*. This fact often escapes the notice of those who profess the Popper's ideal of a researcher who single-handedly grapples with facts by formulating bold hypotheses and making attempts (which may fail) to disprove them. That ideal, illuminating as it may be in some respects, is certainly not universally applicable to the enormous undertaking called science.

It seems therefore – and there is probably nothing strikingly new in it – that those research operations which yield statements at a fairly low level of generality can most easily be subordinated to methods. Such operations are repeatable (which is necessary condition of being covered by a method), because partial research goals are repeatable, too, in the sense of formal identity.

This applies above all to the accumulation and processing of data. By data processing we mean here submitting data to various condensing operations following which such data become clearer, so that their agreement or disagreement with a given hypotheses is revealed. Such role is played, for instance, by the taxonomic procedure which formally consists in reducing the position of a point in a many-dimensional space to the membership of that point in a subset of that space. To put it more plainly, a multifaceted description of elements of a set is replaced by a partition of that set. The value of that method obviously depends on whether it eliminates such information which is superfluous from the point of view of further research. A similar role is played by measurement, which additionally enables us to replace empirical relations which hold among certain objects by arithmetical relations holding among the numbers which are assigned to the classes of such objects.

Methods of data processing are being developed and improved, due especially to present computation techniques. But it must be pointed out here that the very concept of 'data' is incessantly evolving. As science develops, we class as 'data' results of more and more indirect observations, whose connection with the phenomena that are the subject-matter of a given theory is based on the functioning of intricate devices and on our conclusions drawn from general statements, incorporated within a given discipline at an earlier date.

This modifies, and adds to, the set of those methods which are used in establishing of what is called facts, or, more precisely, of what in a given discipline and at a given time is taken to be 'facts'. This is one of the indicators of progress in science: a discovery of new laws enables us to devise new methods of establishing facts. For instance, the laws of modern physics have made it possible to develop the radioactive carbon method, which is used to determine the age of organic substances.

It is worth noting in this connection that the algorithmization of methods of data accumulation and processing also marks progress in science since the precision of these methods enables us to let the computers do the work: computers carry out the appropriate sequences of physical and computational operations and yield results in the form of processed data, e.g., correlation matrices, best-fitted curves, etc. The time which researchers formerly had to spend on such operations can now be used by them for creative activity.

This will suffice to demonstrate (or perhaps to recall) that science at its lower levels is characterized by the use of various methods which are rational in Ajdukiewicz's sense, that is, they are effective in attaining those goals which mark initial stage of research. Is singling out and justification of such methods the task of a methodologist? Undoubtedly so. The results of such investigations usually are included in lectures and monographs on the methodologies of the various disciplines, since they are more specialized than statements formulated in general methodology.

The latter, however, is not confined to an analysis of data accumulation and processing, and rather concentrates on something else. It cannot leave outside the sphere of its interests that part of research which is carried out at the highest levels of generality as it consists in explaining established facts and hypothetical general statements. Let that part of research be called, for brevity, theory construction.

Are there methods of theory construction? The ambiguity of this formulation has often led to misunderstandings, as mentioned at the beginning of this paper. By posing such a question we may mean two quite different problems: (a) are there methods (in the sense of repeatable procedures) of discovering theories? (b) are there methods (conceiving in the same way) of justification of theories?

This distinction, which is probably due to Hans Reichenbach,[2] takes into consideration the self-evident fact that justification of theories is subject to public inspection and hence must satisfy certain conditions which are stated explicitly or at least implicitly accepted by the community of scientists. We may, therefore, speak about methods of justification of theories at least in the sense that such methods codify certain generally available way of finding out whether the theorems advanced by a given theory are acceptable or not.

It would be hard to overestimate the importance of methods of justification of theories in science. To put it briefly, they protect science against error by ensuring intersubjective verifiability of statements made in scientific theories. The criterion of the consensus of competent people, which is accepted as the regulator of science (Ziman in his book raises it to the status of the principal scientific *method*) can work precisely owing to the existence of commonly accepted methods of justification of theories. A discovery is unique by its very nature since it depends on imagination, inventiveness and other similar but fairly rare traits of a scientist's personality. Testing of a discovery is egalitarian: it can be carried out by any scientist who is competent in a given field, if he knows the methods of verification of theories.

In order to bar possible misunderstandings, we have to explain that verification usually consists in both repeating and checking the justification already achieved. For instance, we carry out an experiment again, or check the steps of an intricate process of inference that leads from the assumptions of a given theory to statements about facts. Such an experiment may originally require great ingenuity; the same can be said about such a process of inference. Thus, when we speak about methods of justification of theories we do not mean those ways which make justification possible or easier. We mean the way of ascertaining that a given theory *is* justified.

All this is, of course, greatly simplified. Except for the deductive sciences, justification of theorems is never complete, which leaves a margin of freedom for assessing the value of arguments, and for deciding whether the degree of justification is sufficiently high. Hence the controversial nature of those theorems which are at the border line of justification which most scientists would accept as sufficient.

Moreover, the earlier comment on non-verbalized and even unconsciously followed procedures is fully applicable to the methods of justification. That is exactly the most difficult task of the methodology of science: to reconstruct the standard principles of theory construction, of introducing new concepts into a theory, of basing hypotheses on experimental data, etc., in a word, to bring out and explicitly to formulate all that by which scientists are *in fact* guided when they accept a given theorem as justified. It must be said that we are still a long distance away from attaining the ideal of full and sufficiently precise knowledge in that field. Models of theories constructed so far either refer to small parts of the body of science or are so general that their applicability to existing theories is doubtful. This refers, of course, to the empirical disciplines, since the deductive sciences are in a much better position when it comes to the methods of substantiation used in them.

As we can see, the question about the methods of theory construction can be answered in the affirmative on the condition that what we have in mind are methods of justification. On the other hand, the issue of methods of making discoveries raises doubts which were the starting point of our reflections. This is why the expression 'a method of discovery' seems to be a contradiction in terms.

But, as someone might object, creative thinking is a subject of various investigations. There is e.g. heuristics, that is, the discipline which formulates certain recommendations in that connection. Can these recommendations be regarded as methods of making discoveries?

The answer obviously depends on how broadly we interpret the term *method*. Consider, for instance, one of the most frequently mentioned heuristic recommendations: if you cannot find the solution of a problem for a long time, stop thinking about it, busy yourself with something else, and then come back to your problem. This advice is supported by examples drawn from the history of science and also by the supposition that the search for the solution is continued on the level of subconsciousness and at one moment may result in the discovery which comes unexpectedly and is experienced as a sudden revelation. Such a process may in fact be impeded by repeated but failed efforts as these obstruct the unconscious search for completely new paths of thought.

This seems to be a good advice. And yet the present author would suggest that it cannot be called a *method* of discovering anything. The arguments against calling that a method are both the imprecision of the recommendation (when is one to stop thinking about his problem? For how long? What is he to do in the meantime? How many times is he to repeat that procedure?) and the impossibility of measuring or even estimating its effectiveness. It is obvious that success depends here mainly on the difficulty of the problem and the specific abilities of the person involved.

This is so probably in the case of most, if not all, heuristic recommendations. They are certainly useful, and the refusal to treat them as methods is not intended to discredit them. The point is merely to indicate the essential difference between such recommendations and fairly precise instructions which can be proved effective with regard to a given goal (e.g., methods of statistical inference, methods of measurement, etc.). It seems reasonable not to endow the term *method* with too many meanings, since it may lose all its sense if used indiscriminately. Like all terminological proposals, the present one is also to be assessed from the point of view of convenience.

I am inclined to see the value of heuristic recommendations in their answering the question: how to provide conditions that are conducive to making discoveries? The problem is of considerable practical importance, but it must be treated in a highly individual manner. It depends on a person's temper, whether his creative thinking is assisted by silence prevailing in a sound-proof studio, or by the noise of a coffee-house; by his sitting motionless or walking; by his following a strict schedule of work or by admitting a certain lack of order in his work.

The heuristic of better more advanced, e.g., mathematics, offers much more specific recommendations, more strictly connected with the subject matter of research.[3] It points out the ways of finding and utilizing formal analogies, of making tentative generalizations, and of proving the regularities discovered. These and other, still more detailed, recommendations are certainly of valuable assistance in the search for solutions (and sometimes in the very formulation of problems), the more so as publications merely report the results obtained, and do not show how ideas are born, even so far as that can be shown at all. But heuristics can hardly be expected to provide recipes for making discoveries.

A fortiori, the same cannot be expected from the methodology of sciences, which is concerned with something else. Disappointment with general methodology, which is often voiced by scientists, is usually due to the fact that they expect from it methods of discovery, but they are given methods of justification instead. These they find of little use, since they very well know how to justify theorems and theories, and moreover, the understanding of these methods is far from easy since it requires the mastery of specialized concepts and comprehension of rather pedantic distinctions. The study of general methodology (as the science of techniques of justification and the rationality of such techniques), as distinct from the methodology of the various disciplines (interpreted as the science of accumulation and processing of data) is suggested by considerations other than direct practical advantages to be derived by individual researches. To put it briefly, those considerations are ones of *Weltanschauung*. It is, therefore, not a coincidence that the term *philosophy of science*, used in the English-language literature of the subject, semantically comes very close to what present writer suggests to call general methodology of sciences.

Our reflections have focussed on the oppositions between inventiveness and routine in science. We might ask what is the proportion in which these two ingredients contribute to research activity.

A clear-cut answer is practically impossible because – as far as I know – the problem has not been studied by anyone. It may be supposed that the proportion varies from discipline to discipline. For instance, a larger share of inventiveness might be expected in those sciences which more often employ deduction in the strict sense of the term: in these disciplines which are less 'theory-permeated' and are more descriptive research follows to a great extent on standard procedures of collecting, processing and interpreting empirical data. Yet even the selection of empirical data is hardly ever mechanical only: much ingenuity is sometimes required to associate a given problem with an already known method that has been tested in a wholly different field. Nevertheless I am inclined to think that the share of inventiveness in the whole of research activities is usually overestimated.

There is nothing strange in such an attitude. When we think of science we most often mean the standards set by Newton or Pasteur. And in a sense we are right in doing so since that is science in its most admirable form and it should be, I think, treated as *the* standard. But all research which is more routine-based and is being carried out by hosts of scientific workers also *is* science.

This statement is not an apology of mediocrity in science, since it does not valuate anything: it just describes what seems to be the actual state of affairs in science.

Let us return to the questions posed at the beginning. The concept of method as applied to research seems paradoxical if a method is interpreted as an algorithm (or something that comes very close to an algorithm), and if science is interpreted as a purely creative activity. Yet neither methods are always fully defined, nor does research consist only in creative thinking.

Translated by *Olgierd Wojtasiewicz*

NOTES

[1] The realization of that fact is now fairly common. See in this connection J. Ziman, *Public Knowledge*, Cambridge University Press 1968.

[2] Cf. H. Reichenbach, *The Rise of Scientific Philosophy*, 1951.

[3] See G. Polya, *How To Solve It*? Princeton University Press 1945, and by the same author, *Mathematical Discovery. On Understanding, Learning and Teaching Problem Solving* vol. 1 1962, vol. 2 1965.

4
SOCIOLOGY AND MODELS OF RATIONAL BEHAVIOUR

INTRODUCTION

In one of its many meanings, the word 'model' denotes an ideal, rarely (if ever) attained. Thus we speak, e.g. of 'model behaviour'. It is rather important to distinguish, in methodological context, this meaning of 'model' from the more common, descriptive one. The difference is essentially that of purpose.

When constructing a descriptive model of a given kind of phenomena, the scientist aims at mirroring some of this properties. And the model is judged accordingly: the more faithfully it represents the properties of the real system, the better it is. The criterion used in the appraisal of normative models (this is how I propose to refer to them) is fundamentally different. They are meant to express value judgments, rather than statement of fact. Their role consists in making the evaluative concepts more precise, in investigating their implied properties and solving the problem of their consistency. Normative models are judged satisfactory if they perform this function well.

A large family of normative models aim at elucidation of the concept of rational behaviour, in its different meanings and relative to various types of human activity. Naturally enough, models of rational behaviour are not uninteresting from the point of view of sociology. Indirectly, they serve descriptive purpose. Even if it is argued that people never behave quite rationally, some concept of rationality is needed for analysing actual behaviour. Typical questions asked in this context are: How people depart from rationality? What influences them to do so? Under which conditions? etc. In order to speak meaningfully of irrationality one must have an idea of what 'being rational' means.

Logic is normative in a somewhat similar sense: it provides a model of consistency in thinking. No one would seriously maintain that anyone's body of beliefs is closed with respect to the relation of logical consequence and contains no inconsistent beliefs. The notion of logical consequence is not rejected for that reason.

Pushing this simile a step further, one might ask, what about the plurality of logical systems? Granted that classical two-valued logic is predominantly in use, there are still adherents of other systems, e.g. the constructivists in mathematics, believers in the necessity of three-valued logic in quantum mechanics, etc. Is the same valid for models of rational behaviour?

The answer, of course, is that in legislating upon behaviour we have still more freedom. Our intuitive requirements for the rationality of behaviour are obviously less precise than those for rational reasoning. There exist competing concepts of rational behaviour, some of them essentially different from one another. It is impossible to prove the absolute superiority of one such concept over its alternatives. The most that can be done is to make their properties explicit. This type of elucidation is achieved by means of formal models.

Such models are not universal in the sense of covering all possible types of behaviour. They apply only to cases which satisfy the presuppositions, explicitly stated for each model. A well-known distinction is that between decision making proper (i.e. when the uncertainty is not attributable to conscious decisions of some human agent) and games (i.e. decision making in situations of conflict, partial; or total, with an intelligent opponent). Many other differentiations exist.

In the following remarks I shall be concerned with normative models mainly from the point of view of their role, actual and potential, in sociology and in sociological applications (sometimes referred to as social engineering or sociotechnics). The discussion will be centered around four topics: (1) The normative models of individual choice behaviour; (2) The models of aggregating individual choice into social choice; (3) The models of behaviour in conflict; (4) The role of normative models in sociotechnics.

My aim being to indicate main points only, I shall not attempt to be exhaustive. This also explains somewhat haphazard bibliographical references. On the other hand, I will try to stress some less well known tendencies.

The approach to human action via normative models is not advocated here as a kind of panacea. On the contrary, its limitations will be pointed out. It seems, however, that it does not get sufficient attention as a type of approach different from the usually adopted, i.e. descriptive one. If such is the case, the present discussion is perhaps justified.

Some mention of the formal properties of the models is unavoidable. It will be kept down to a minimum. For details, couched in appropriate symbolism, the reader is referred to other sources, where rigorous definitions and proofs of theorems can be found.

1 MODELLING INDIVIDUAL CHOICE BEHAVIOUR

One of the difficulties in decision making, known from everyday experience, is the uncertainly concerning the consequences of contemplated actions. A large group of normative models deals with the problem, what does it mean to behave rationally in the face of uncertainty.

The prevalent paradigm, some 25 years old, consist in splitting the factors conditioning the future into two groups. The variables under decision maker's control are amalgamated (by means of the Cartesian product of the values they can assume) into two classes: first of them, say *A*, consists of total descriptions of decision maker's possible ways of behaviour. These are called strategies or acts; it is from the set A that the decision maker is assumed to choose. The variables which the decision maker cannot (or does not know how to) control are handled in a similar way. This result in a set, say *S*, of so-called states of the world, i.e. complete descriptions of possible conditions under which the decision making takes place.

The sense in which both *A* and *S* are 'complete' is the following: any combination of a in *A* and *s i S* is sufficient to determine all the consequences that matter to the agent at the moment of his decision. It follows that the elements of the Cartesian product $A \times S$ constitute, for the decision maker, an exhaustive set of possibilities: he need not bother about any other feature of the world.

To this powerful assumption another one is usually added, no less strong. It is assumed that the agent's valuations over the set $A \times S$ are expressed by one single numerical function *u*, called utility. In fact, however, the agent evaluates the elements of $A \times S$ from several points of view, conflicting more often than not. Postponing comment, let us briefly describe the function the model $\langle A, S, u\rangle$ is intended to perform.

First and foremost, it makes it possible to define with sufficient precision the concept of rational behaviour under uncertainty. There is more than one way of doing this. Without going into details, two basically different approaches ought to be distinguished. One (the so-called Bayesian solution) introduces a probability distribution *p* over the set *S* of states of the world. The probabilities represent the subject's personal estimates of chances of the elements of *S* to happen. Rational behaviour then consists in choosing this strategy from *A* which maximizes the expected value of *u* with respect to *p*.

The other, 'objectivist' concept of rationality postulates that the agent choose according to some criterion based exclusively on $\langle A, S, u\rangle$ and possessing certain desirable properties. Several such criteria have been suggested (maximin, minimax loss, etc.), some of them parametric (e.g. Hurwicz's α), with the understanding that the decision maker fixes the value of the parameter in the way which best represents his attitude (e.g. his 'degree of pessimism' in the case of α).

The advocated concept of rational behaviour, be it Bayesian or some objectivist criterion, carries over to the problem of handling information. For a social scientist the problem is not unimportant, at least insofar as he makes inferences from sample to population. Which rule of inference is adopted, depends on the concept of rational choice in conditions of uncertainty.

The difference between the Bayesian and non-Bayesian approach is quite deep, at least in interpretation if not in the final outcome. It expresses the difference between two possible philosophies of rational behaviour. The Bayesian philosophy reduces rationality to consistency. The subject may be a fool in his assignment of probabilities to states of the world: he nonetheless remains rational as long as he manages to be consistent in his calculus of chances.

The non-Bayesian approach tries to exclude the subjective element as far as possible: once the decision problem $\langle A, S, u\rangle$ is given, its solution automatically follows from the criterion adopted; it could, in principle, be found by a computer. A 'foolish' estimate of chances is here impossible. There is a price to be paid for this objectivity: genuine knowledge about the relative likelihood of states of the world cannot be expressed in non-Bayesian terms. Uncertainty is here interpreted as complete ignorance.

This is not the place to discuss further the relative merits of the two concepts of rationality. Of some interest from the sociological point of view is the problem, how the adoption of the first or the second concept influences the notion of responsibility, particularly social responsibility, for the decision made. The Bayesian concept of rationality introduces explicitly an uncontrollable element, viz. subjective estimate of chances. Hence the difficulty of deciding whether the decision was correct or not (if the notion of correctness is not to be defined simply in term of the outcome actually obtained). Within the Bayesian framework, such questions hardly make sense, apart from consistency in the utility (probability calculus). The way is, therefore, open to well-known deviations from rationality, e.g. assigning higher probabilities to more desirable events (wishful thinking).

The 'objectivist' solution is free from this objection. Once the criterion of decision making has been agreed upon, the correctness or incorrectness of a particular decision, given $\langle A, S, u\rangle$, is easy to establish. On the other hand, it may be thought that solving decision problems in such a way lacks the necessary elasticity, leaving no place for background knowledge, intuition, etc.

The model assumes the function u as given, thus avoiding the difficult and subtle problem: what is the rational way of solving a conflict of values? It is more than likely that a general answer to this question does no exist. However, the existence of partial solutions cannot be excluded. Research into the actual process of amalgamating many-dimensional valuations into a single utility function (or at least a preference ordering) is needed if the normative model is not to be far removed from reality. This is a task for psychology and possibly social psychology, insofar as the process of arriving at the 'overall' valuation is socially conditioned.

By itself, any model of rational decision is probably more important in psychology or economics than in sociology. However, its paradigm is applied in other contexts, such as collective decision making. Also, any analysis of the role of the social scientist, as an expert, in actual decision making presupposes some type of normative model. We shall, therefore, come back in the last section to some problems discussed above.

2 MODELLING COLLECTIVE CHOICE BEHAVIOUR

In the classical version, the problem originated probably with Concorcet. The articulation leading to the famous 'paradox of the welfare function' is due to

Arrow.[1] In broad outline, the problem can be stated as follows.

Given a set of agents ('society') and a set A of alternatives the society faces, we look for a function transforming any configuration of individual preference orderings into a 'social' preference ordering. The function ought to satisfy certain requirements, embodying our concept of rationality in accommodating individual preferences – a typically normative model. The paradoxical result, obtained by Arrow, shows that there is no welfare function satisfying an intuitively appealing set of postulates. Later on, it has been proved[2] that the paradox remains if we replace preferences by choices, which seems more in accordance with the social practice (voting procedures, for instance).

The fact that the notion of socially rational amalgamation of preferences (or choices) is self-contradictory is instructive in itself. It constitutes a good example of an unexpected result obtained by means of a normative, formal model.

There is no need to stress the social aspects of the 'welfare function' problem: it asks for a definition of rationality of collective decision. It is not surprising that Arrow's paradox, both elegant and general, initiated a series of studies directed at bypassing the newly found impossibility. It has been suggested either to relax the condition imposed on the welfare function or to restrict its domain, on the ground that actually existing societies are homogeneous to a degree which excludes the variation of tastes leading to the paradox. The literature is impressive and still growing.[3]

At the same time, the model's very generality is the cause of some dissatisfaction. It is objected against the assumptions of the model that they are too far removed from the reality of human societies actually existing. As Radner[4] puts it, 'it would seem that very definition of group rationality may have to depend on the particular alternatives available to each group member, on each member's information about the environment and about the other members of the group, on the possibilities for communication among group members, and on the degree and nature of the differences among the ways different group members order the consequences of group action'.

A significant step towards more realistic (and more complex) models of rational group behaviour has recently been made by Coleman.[5] The new elements he introduced can be described as follows.

It is seldom the case that actual collective decision are made in isolation. Therefore, Coleman's model explicitly assumes that the group faces a series of problems. Each member of the group has a distribution of interest (utility) over the set of problems, measuring the relative importance, for him, of the problem being solved in a particular way. Also (this is another new assumption) he has a power distribution over the problems, measuring his relative influence on their solution. Both measures are normalized, summing up to 1.

The basic idea now is that if the members of the group are allowed to trade off power over one decision problem for power over another one, there ought to be an optimal overall outcome, in the sense of equilibrium of power and satisfaction obtainable under the circumstances. A relatively simple calculus is developed for the purpose of finding this equilibrium, based on multiplication of the two matrices: the matrix of interest and that of power.

Although Coleman's model has been constructed for descriptive purposes, it has an obvious normative interpretation: given the matrices of interest and power, what would be the rational way for the members of the group to behave, if each individual's only aim is to maximize his own utility function? In the games specially devised to accommodate the assumptions of the model, the actual behaviour of the players exhibited a good fit with the model's predictions. This would show that they behaved rationally, in the sense defined above.

Coleman's model, though much nearer social reality than the classical one, is still not easily applicable in situations which have not been artificially created. The main obstacle is, of course, the measure of an individual's power over a given decision problem (there are some techniques for measuring desirability). Without being able to measure power, we cannot operationally speak of its exchange. Consequently, it is impossible to tell whether the final distribution of power confirms or falsifies the hypothesis that the collective behaviour has been rational.

In spite of the progress that has been made thanks to Coleman and his school, the problem of rationality in collective choice behaviour is still largely unexplored. To give only one example: the great variety of formal procedures regulating the process of collective decision making remains to be analysed in terms of normative models: do they satisfy the desiderata of such models? given the procedures, what a rational individual's behaviour ought to be?

There is little doubt that models of collective decision making are of primary importance in social science, since they represent the constitutive aspects of society, in its idealized form.

3 MODELLING CONFLICT

We owe it to the game theory that the notion of conflict has been expressed quite generally and with sufficient precision. Game-theoretic formalism made it possible to ask for a definition of rational behaviour in situations of conflict. It proved to be a question incapable of unique and general solution, except for the very special case of two-person total conflict (in the sense that their valuations of outcomes are exactly opposite). The case last mentioned belongs to the earliest formal models of rational behaviour.

The usefulness of game-theoretic models in social science has been the subject of much debate. The hopes to use them directly as predictors of social behaviour have not been satisfied, with the possible exception of some very simple cases of complete antagonism. This was found disappointing and led to criticism of game-theoretic models, interpreted as a description of actual behaviour. As normative models, however, they remain useful if only because they clarify basic notions relating to interpersonal conflict. The field of potential applications is rather wide: members of any group are at least partially in conflict.

However, as in the case of Arrow's model of collective decision, the generality of game theory is the source of its weakness. It is difficult to see how most real-life situations can be adequately translated into the very general, and for this reason somewhat poor, language of game theory. The way from the theory to reality appears to be long and uncertain.

One obvious remedy is to try to construct conceptually richer models, adapted to special classes of actual situations of conflict and therefore more easily applicable. This seems to be the present-day trend of research. An instance of it are Coleman's 'exchange of power' games, described above. I will mention a few other examples, to exhibit the potential usefulness of this approach in social science.

One of them, the so-called generalized prisoner's dilemma,[6] can be described as follows. In an *n*-person society, each individual has two alternatives: either to behave cooperatively or selfishly. The payoff to each member of society is a strictly increasing function of the number of people who chose cooperative behaviour. But the 'selfish' people get in addition some bonus which constitutes an incentive to defect. A suitable manipulation of the parameters of the model, together with assumptions concerning the knowledge about other people's behaviour, make the rational choice vary.

An interesting feature of the above model can be stated thus. It is plausible to assume that the choice of cooperative strategy is desirable from the society' point of view. Then the model makes it possible to formulate and solve the problem, how to create conditions which would force a completely egoistical individual to behave in the socially desirable manner.

The same purpose is served by the classical, and perhaps better known, 'games of fair division'. Here again, the model is so constructed that the rational behaviour (in the sense of maximization of utility) would consist in furthering socially desirable aim. In this case, the aim consists in equal division of some divisible good. The simplest is the two person variant: you divide, I choose. The rule has been generalized to *n*-person game.[7]

Still another example[8] of the same type of problem is connected with the passing of information between players. Suppose *A* can convincingly (perhaps through an intermediary) pass to *B* information concerning his own choice. The message need not be true, on the contrary: it may be misleading on purpose. Under what structure of the game rationality would force *A* to tell the truth? In other words, how to fashion social conditions in order that 'honesty be the best policy'?

Of course, none of the models mentioned above pretends to solve the most difficult problem of influencing the individual by means of a suitable social structure. The game in question has a very narrow field of applications. Also, it is based on the assumption of rationality in the sense of consistent selfishness (at least, if the payoffs are interpreted as money or its equivalent; if they are interpreted as utilities, the problem of measurement arises). Whether this assumption is approximated by actual behaviour, is an open problem in any particular application. Nevertheless, such models are laboratory cases of building up an efficient social structure. This is what makes them interesting from the sociological point of view.

In terms of game-theoretic formalism, it is possible to define certain concepts which are widely used in social science, although their meaning is somewhat vague. Strategic power, blackmail, degree of antagonism inherent in the conflict – are examples.

Finally, it seems that some classical concepts remain unexploited, although they explicitly leave room for sociological interpretation. One of them is the concept of ψ-stability. In n-person game theory, the function ψ imposes a restriction on the process of coalition formation. A sensible way to interpret this restriction would be to say that certain transitions from one coalition structure to another are impossible on social grounds. It remains for a social scientist to specify this kind of assumptions for n-person society and to investigate their influence on the ψ function.

To close this section, I would like to reiterate that game-theoretic models explicitly assume the rationality of the players. Once this assumption is suspended, even the best-established solutions are no longer valid. In a completely antagonistic situation, it would not be sensible to choose one's strategy according to maximin if it were known about the opponent that he believed in unlucky numbers and will be influenced in his choice of strategy by the numbers that have been assigned to them. Models constructed by game theory are classical examples of normative approach to human behaviour.

4 PARTICIPATING IN DECISION MAKING

What help can the social scientist expect from models of rational behaviour when he advises a decision maker on how to make a rational choice? Within the framework of individual decision making model, the answer seems simple enough. Given the decision problem $\langle A, S, u \rangle$, the only point on which the decision maker needs to be advised is the actual state of the world. This is, therefore, the question which the expert is supposed to answer: which element of S actually obtains (diagnosis) or will obtain in some specified future (prognosis).

The time variable seldom appears in the formal apparatus of the model, which obscures the difference between diagnosis and prognosis. It does exert, however, and indirect influence.

In the case of diagnosis, when the elements of A (decision maker's actions) and those of S (states of the world) are approximately simultaneous, the expert's answer to the question concerning S may have the form of probability distribution on S, ideally a zero-one distribution.

On the other hand if, the elements of S describe the world at some future time, whereas the choice from A must be taken now, the choice from A becomes one of the elements conditioning the states of the world. It follows that prognosis, in the form of probability distribution on S, depends on the a chosen. Assuming that the prognosis is presented to the decision maker before he makes his choice (indeed it is to help him in his decision), it must be probabilistic and conditional at the same time. The prognosis is then the conjunction, for all a in A, of statements having the form: if you do a then the probability distribution on S is p_s.

Formally, therefore, the prognosis is the matrix of conditional probabilities, of the type Jeffrey[9] has advocated for decision models, instead of one probability distribution. The entropy of the distributions is, as a rule, the higher the more removed in time from its object the prognosis is.

The situation is basically unchanged if additional information is sought before the expert presents his advice. Typically, the information consists in singling out an element of a set, say X, connected probabilitywise with S. This induces an *a posteriori* distribution on S, or (in terms of u) a new risk associated with actions in A. If the adopted model of decision making is of the non-Bayesian type (cf. section 1) the character and role of information is somewhat more complex. We will leave this case out, on the ground that non-Bayesian models are seldom used in the cooperation between expert and decision maker.

Thus far, the role of expert is clear-cut and comparatively simple. Under such conditions, the social scientist's task is like that of any other expert. It is not often, however, that his advice is asked after the decision problem has been unequivocally formulated, up to the uncertainty about S. Typically, his role in the process of decision making consists in helping to shape the problem itself. The problem of influencing social reality is, as a rule, much less clear-cut than, say, making decisions on some technical matter: the decision maker not only knows less but is also less certain about his own valuations.

What is the role of expert in formulating the decision problem? It is tempting to say that he brings in his knowledge of facts, whereas decision maker supplies the value judgments. In terms of the model, expert defines the sets A and S, thus telling the decision maker what are the options, and the latter can define the function u, to be maximized. In reality the process is far from being as simple as that.

The reason is that statements of fact cannot, in this context, be separated from value judgment. It is not possible to say what can be done if some rough idea of the objective has not been previously given. And the same goes for the consequences: their definition presupposes that the expert can tell what is important from the decision maker's point of view and what is not. Vice versa, having learned from the expert about his possible actions and their consequences, the decision maker will, in general, have to change his valuations, because e.g. he learned about some undesirable consequences of the contemplated behaviour.

It follows that the process of arriving at a satisfactory formulation of the decision problem consists in repeated use of feed-back between the two parties, each stage bringing them somewhat closer to the adequate representation of the choice the decision maker faces.

The formal model helps in this process as an ideal structure of what is sought. Does the final product of the cooperation actually have the form of $\langle A, S, u \rangle$, with an estimate of chances, say p or $\{p_a\}_{a \in A}$, on S supplied by the expert? Probably not very often. If the problem is as well defined as that, and the decision maker accepts his adviser's estimate of probability, then there is nothing left for him to do. The decision follows automatically through maximization of expected utility. The only possible difficulties are of computational nature. Observation suggests that with most non-trivial decision problems this is not the case.

The main obstacle is, of course, that the valuations of the elements of $A \times S$ do not conveniently sum up to one overall utility. Not only are they conflicting and hardly expressible in a common unit so that a balance could be drawn, but they usually cannot be treated as utilities, at least in the following sense: under conditions of uncertainty, utilities reduce to one single number, viz. their expected value. Payoffs in money do not always have the last mentioned property. The value of x monetary units with probability p or y units with probability $1 - p$ is not necessarily $xp + y(1 - p)$. It may be more or less than that, according to the decision maker's attitude towards risk.

Risk taking is something that the decision maker cannot be relieved from. In the formal model of rational behaviour no mention is made of risk, simply because it is assumed that valuations, in the form of utility function, have the convenient property of linearity. In applications, it is difficult to speak in terms of utilities.

Probably, therefore, the furthest that the decision maker can be guided by his adviser is this. Suppose that the aspects under which he evaluates the consequences of his actions are numbered from 1 to n. Let $I = \{1, 2, \ldots, n\}$. Then the result of his cooperation with the expert can be represented as

$$\langle A, S, \{u_i\}_{i \in I}, \{p_a\}_{a \in A} \rangle \qquad (*)$$

In the above expression, u_i is the evaluation of the consequences under their i-th aspect (say, the number of the people getting employment, or the pollution of atmosphere, or financial expenditure, etc.). If the function u_i have the properties of utility the decision problem (*) can be simplified to

$$\langle A, I, \{u_i^*\}_{i \in I} \rangle \qquad (**)$$

by means of

$$u_i^*(a) =_{df} \textstyle\sum_s u_i(a, s)p_a(s)$$

In other words, the uncertainty is eliminated, due to the estimate of probabilities, conditional upon a. What remains for the decision maker to do is to establish a preference ordering on A, based on the position of A's elements in the n-space of partial valuations.

Theories of rational behaviour have nothing to say concerning the choice of a in (*) and very little about that in (**). It would maintain, however, that this is no reason to think little of them. They bring at least the above-described ordering into the complexity of decision problems usually met with in applied social science. If the adviser can take the decision, maker thus far through the preparatory stages of decision, he makes him realize what can be done, what are the possible consequences, what are the risk to take under any aspect considered. This helps, in general, to improve decision.

5 CONCLUDING REMARKS

It ought to be clear from the above discussion that models of individual rational behaviour, even if not always applicable literally, serve as a guiding principle in the process of making the decision emerge out of the initial chaos of vague objectives and mostly irrelevant data. Models of collective behaviour and of rational decision in conflict do the same for acting in explicitly defined social contexts.

One limitation ought perhaps to be mentioned here. In all models of rational behaviour, it has been assumed that the complete set of alternative actions was well defined. On this ground, the problem was that of choice. Whereas, it might be maintained that the most desirable way to solve a problem of the type, say, $\langle A, S, u\rangle$ would consist in adding to the set A such an action, previously not taken into account, which would be at least as good as any action in A, in each of the possible worlds. Such a solution, let us call it 'creative', would undoubtedly be the most satisfactory way to deal with the uncertainty: the decision maker would (at least in this context) have no need to know which state of the world actually obtains. Creative solutions are, by their very nature, not formalizable.

If the normative models are not intended directly to describe human behaviour, what are the criteria for evaluating them? First and foremost, a model of this type must correctly express at least some features of the usually vague and sometimes inconsistent ideal we have in mind while constructing them. This is essentially a matter of judgment: a rigorous proof of adequacy would be impossible. Another desirable property, perhaps not necessary as the first one, relates to the assumptions of the model. They ought to be realistic, i.e. approximated by the actual conditions of the behaviour under discussion. If they are not, we have no means of telling whether the behaviour is rational or not, in the sense postulated by the model.

The formal models we have been concerned with are intended to cover only one meaning of rationality of behaviour. Another one, possibly more common, refers to the soundness of beliefs on which the behaviour is based. Irrational in this sense are beliefs entertained without sufficient grounds or even contrary to existing evidence. Indirectly, behaviour based on such beliefs would also be called irrational.

Rationality in this sense is, of course, relative to the meaning of 'sufficient justification of a belief'. As this last concept changes with the progress of knowledge (logic and methodology included), so does the concept of rationality.

Rationality, as applied to the body of beliefs, was irrelevant to the above discussion, insofar as the models considered were, in a sense, hypothetical: if the description of the decision maker's choice situation was correct then this is what he ought to do. Insofar as the decision maker cares about the actual outcome of his decision, he also has to be rational in his beliefs. In this sense, the two concepts of rational behaviour are complementary.

Finally, one might add that there are normative models other than those which aim at characterizing rationality, in its various meanings. There is a large family of models trying to describe 'justice' in certain well-defined contexts. The model of collective decision by Arrow partly belongs to them, since it includes the postulate of positive association of the welfare function with all individual preferences, and certain similar desiderata. Another example is the formalized notion of a 'just compromise' in non-zero-sum games. Still another is a rule of distribution of some non-divisible goods, variously valued by the members of society.[10] They are outside the scope of the present paper.

REFERENCES

1. Kenneth J. Arrow, *Social Choice and Individual Values*, J. Wiley, 1951.
2. Bengt Hanson, 'Choice Structures and Preferences Relations', (*Synthese* 18, 1968), also: 'Voting and Group Decision Function', (*Synthese* 20, 1969).
3. One recent example, a special issue of the journal *Theory and Decision* (September 1973) entitled: 'Arrow's Paradox and Collective Choice'.
4. Roy Radner, 'Normative Theory of Individual Decision. An Introduction', in: *Decision and Organization*, ed. by C. B. McGuire and Roy Radner, North Holland, 1972.
5. James S. Coleman, *The Mathematics of Collective Action*, 1973.
6. See, for instance, T. R. Burns and L. D. Meeker, 'Structural Properties and Resolutions of the Prisoner's Dilemma Game', in: *Game Theory as a Theory of Conflict Resolution*, ed. by A. Rapoport, D. Reidel, 1974. Also J. Grzelak and T. Tyszka, 'Competition and Cooperation in Conflict of Interests' (in Polish), *Studia Psychologiczne*, 1974, p. 145–179.
7. Hugo Steinhaus, 'The Problem of Fair Division', *Econometrica*, 16, 1948.
8. Klemens Szaniawski, 'On the Value of Information in Games', *Proceedings of the CESMAP Seminar on Decision Theory*, Beaulie-Ste-Assise, Dec. 1973.
9. Richard Jeffrey, *The Logic of Decision*, McGraw-Hill, 1965.
10. Klemens Szaniawski, 'The Concept of Distribution of Goods', *Studia Filozoficzne*, 1970. See this volume, paper 15, pp. 128–139.

5

MATHEMATICAL MODELS AND SOCIAL FACTS*

There is no theoretical trend in sociology that could be termed 'mathematism' by analogy to functionalism, historicism, and structuralism. The term 'mathematical sociology' is becoming more frequent, but it is precisely the different character of that term which indicates that something else is involved here than in the case of the 'isms' mentioned above.

The essential difference is in that the classical theoretical trends in sociology advance certain statements about social facts, and substantiate the recommendations to the researchers (e.g. recommendations concerning the way of explaining observed facts) by reference to those very general statements.

Nothing of that kind takes place when mathematical methods are used in sociology. Mathematical sociology is not specifically linked with any opinion about social facts. It is true that such an opinion may influence the choice of a formal model (for instance, cybernetic models are very convenient in the case of studies inspired by functionalism), but no decision about the choice of formal apparatus places the researcher outside mathematical sociology because mathematical sociology does not predetermine anything. To put it briefly, in sociology, as in the case of other disciplines, mathematics is neutral to factual problem: the fact that mathematical models are constructed does not by itself state anything about the standpoint taken by the researcher with respect to the basic issues of the discipline he is concerned with.

This trivial truth is recalled here in order to ward off any misunderstandings that might result from the inclusion of the present paper in a set of studies which discuss the main trends in present-day sociology.

The use of the term 'model' in the title of this paper also deserves a brief comment. This term will be used here in the sense usually ascribed to it in discussions on the application of mathematics: a mathematical model of a given object is tantamount to a set of postulates, formulated in the language of mathematics, which describe that object. Thus it is the other way round than in the case of logic, where the assumptions of a theory (i.e., such an object, in the set-theoretical sense of the term, with respect of which the statements of that theory are true) are satisfied by a model (of the theory). In view of the linguistic usage current in the methodology of the social sciences it would be inconvenient to use the logical concept of model.

For the reasons to be explained below, models, in the sense adopted in this paper, are still rare in sociology. This is not to say that applications of mathematics are rare in sociology. For instance, it is common knowledge that concepts and theorems of mathematical statistics occur in everyday research practice of the sociologists. But they are, of course, not mathematical models of the facts investigated by the sociologists.

The advantages of mathematical representations of the facts are generally known and will not be discussed here. More attention will be paid to the way in which mathematical models can be introduced in sociology.

It is possible to consider *a priori* two alternative ways of doing that, depending on whether a sociological theory or empirical data are taken as the point of departure. Thus, the first method of constructing models would be to mathematize existing sociological theories, and the other, to adjust a formal structure to discovered facts in a way largely independent of any theory. Both cases will be discussed here.

Sociological theories are usually formulated in non-formalized language. This makes it often difficult to realize the connections between the theorems of which such theories consist, and also to deduce rigorously observable and measurable consequences of such theorems. Hence a translation of a sociological theory into the language of mathematics and logic seems to be a useful task, provided that it is feasible. Such tasks in fact have been attempted; some examples can be found in Stefan Nowak's paper on 'The Logic of Reductive Systematization of Social and Behavioural Theories.' On a closer investigation, however, the results of such an operation show that the effect is rather trivial, at least from the formal point of view. In most cases the result amount to particular theorems expressed in the form of general categorical statements (so that syllogistics suffices for deduction) or single statements formulated in the language of conditional probability (so that elementary theory of probability suffices).

It is rather difficult to answer the question why more intricate sociological theories, that is, exactly those which could profit most from the bringing out of their formal structure, lend themselves to mathematization in the least degree. It may be due to the imprecision of everyday language and the metaphors to which it resorts, for they encumber the formulations used in those theories so much that they apparently cannot be translated into the language of mathematics without having their sense distorted. Quite often it is due to the fact that a given theory implicitly or explicitly refers to the concept of cause which has no clear mathematical analogue (as is known, that concept is not be found in the formulations of physical theories, even though it plays an essential role in the applications of such theories).

It is possible that there are other reasons which make it difficult to construct such mathematical models which would become mappings of existing sociological theories. The fact remains that it is difficult to point out any formally interesting mathematical model obtained in this way.

Hence the approach which, in a sense, goes in the reverse direction seems more promising: this approach suggests rather that we should construct a model directly on the basis of available data, following the suggestions of generalized sociological knowledge. A model built in this way would next become a subject of a sociological interpretation.

The problem arises, however, what should be the nature of such data. Now it seems that, above all, they would have to admit the possibility of the introduction of variables. It may be said, with little exaggeration, that without variables there is no mathematics. They need not necessarily be numerical variables: for instance, they may range over the set of human beings, as in the case of the well known application of the theory of social relations to the investigation of kinship system. But numerical variables are certainly the most important of all, since arithmetical operations can be performed on them.

The simplest case is that of a binary variable, i.e., such which can assume two values only, selected by convention (e.g., 0 and 1). The introduction of binary variables is based on a dichotomous classification, and this is probably why variables of that kind so often occur in sociological analyses.

But, in view of the fact that the sociologist has to do with diversified data, a dichotomous classification is not subtle enough: more complex classifications are usually indispensable. Superposition of dichotomous classifications is, of course, one of the ways out. It is in fact resorted to, and the concepts required for that purpose have been worked out fairly well (cf. the well-known book by Paul F. Lazarsfeld *The Algebra of Dichotomous Systems*). But that procedure is inconvenient in some respects.

First of all, the number of the variables increases very rapidly. Correlations that hold between them can be established, but, just because of the primitive nature of the variables the matter ends there for all practical purposes. Yet experience shows that a good mathematical model (i.e., such which can be transformed in a non-trivial way) usually has few variables, but such that functional relations hold between them.

Moreover, applying a dichotomous classification by force very often distorts the nature of the data by obliterating differences which are essential from the sociological point of view: a classification into three classes of equal status is not the same, from the sociologist's point of view, as a double dichotomous classification with one empty class. Yet even in the case of trichotomy the introduction of numbers becomes useless (useless, that is, unless it is justified by some additional assumptions), because operations on those numbers are meaningless. The specific nature of binary variables is due to the fact that two numbers satisfy trivially the assumptions of an interval scale. In the case of variables that assume more than two numerical variables each and are based on a classification, the numbers practically play the role of names, and thus occur in a non-specific role.

We may add, even though this objection is less important, that discontinuous variables and the variables which represent a dichotomous classification, are not subject to operations formulated in classical mathematics. In their case mathematical analysis can serve at most to study their variations in time, where time is treated as continuous. But that, obviously, is a different matter.

As can be seen from above, more subtle methods of data processing than those used so far are a prerequisite of successful formulation of more interesting mathematical models. Endeavours have been made to go beyond simple classification procedures. The best known ones are the Guttman scale and the latent structure analysis. They introduce, on the basis of a classification, a variable which (in principle) is continuous and is defined on the investigated set of objects. Both procedures have their shortcomings.

The Guttman scale rarely fits empirical data satisfactorily. Latent structure analysis achieves usually that fit through the concept of probability involved, but the price paid for it is exceptionally high from the point of view of model construction: in latent structure analysis finding of the value which the variable assumes for a given object requires statistical inference and thus is marked with risk, which sometimes is quite high. The verifiability of a model in which such a variable occurs as one of its elements is, of course, highly problematic. It is, therefore, not surprising that latent structure analysis is as a rule used 'for its own sake', and cannot be the first stage of construction of a more complex model. The problem of variables that would be strictly sociological in nature and at the same time would well lend themselves to operations remains open.

Note also that the sociologist cannot expect such a freedom in assigning numerical values to variables which is characteristic of natural science research. This additionally restricts the possibility of verifying a model. It is so because observation of a single system of values, namely that which is actually given, does not usually suffice. As is known, the most convincing test of a model consists in an experimental verification of the predictions deduced from assumptions (taking under consideration their numerical modifications). This obviously requires experiments. Yet it is not possible to change social fact at will in order to see whether that brings about results which comply with the model adopted. Note also that a simulation of actual processes by a computer does not provide the desired verification, since it is itself hypothetical in nature: it is based precisely on the assumption that the procedure programmed in the computer is a good mapping of the facts under investigation.

To sum up, both the qualitative nature of sociological data (as reflected in extended classifications) and the limited possibility of changing them make it quite difficult to construct a mathematical model for them.

Note that what is meant here is a model that goes beyond empirical data, hence it does not refer to an operation which consists only in rendering in mathematical symbolism what has been found, e.g., in presenting a result of a sociometric research as a graph. Such operations happen to be useful, but they are not satisfactory from the theoretical point of view. The strength of a mathematical model consists in the deductive reasoning which that model makes possible and which goes beyond the data observed. Only such models, which are not equivalent to conjunctions of statements about established facts, perform functions which are the most essential for science: explanation and prediction. Stressing one or the other function may be crucial in the choice of a model: as Coleman remarked, stochastic models (as compared with those termed deterministic) serve the purpose of predicting better than that of explaining. But these roles are interconnected, as they both assume that a given model is logically stronger than the data on which it is based.

The above defects of the data which the sociologist has at his disposal can be compensated only by ingenuity in devising formal systems, which must of necessity reach a higher level than that observable in the case of mathematical models constructed in natural science.

But reality lags far behind that requirement: we usually meet with endeavours to adopt models taken over from other disciplines for sociological needs. Since this has not resulted in any remarkable achievements, the role of mathematics in sociology now is not so much to help construct mathematical models, but rather to help carry out the following tasks:

(1) to process observable data (in order to yield statistical characteristics in the sense of parameters of distribution, to measure stochastic relations conceived in various ways, and to provide classifications and typologies);
(2) to derive statistical characteristics of a given population from data about a sample;
(3) to construct concepts (especially those which do not have any directly observable designata, and thus require establishing well-defined connections with observations).

These tasks are important and certainly deserve an improvement of those instruments that would help carrying them out. They are, however, not identical with the goals for which mathematical models of investigated social facts are constructed.

NOTES

* The paper was read at the conference 'The Principles of Theory Making and Research Directives'.

6
SCIENCE AS A SEARCH FOR INFORMATION

Philosophers of science attempt to explain scientific activity as rational behaviour aimed at attaining particular goals. Here, rationally-aimed behaviour is understood to mean behaviour which uses a method which will be effective in achieving a given goal – i.e. that makes it sufficiently likely that the goal will be attained. However, in most scientific activity, both methods and goals are hypothetical, and scholars themselves are infrequently unaware of them. The application of particular methods is neither essential nor in itself sufficient for the conscious attainment of worthwhile results in scientific work. This leads to doubts about the usefulness of methodological analysis, and preference for sociological or historical interpretation of science.

Leaving aside the concepts cited above, this paper is an attempt to interpret science as a process of searching for information. It is accepted that principles governing the choice of hypotheses are means of obtaining information. These problems are formulated in the categories of the well-known theoretical/decision-making model.

What philosophers of science try to do in their various ways is to understand science as a rational pursuit of certain goals. 'Rational' in the sense that the methods used are effective with respect to a certain goal, i.e. methods that lead to it with sufficient probability. To quote from a 1958 paper by K. Ajdukiewicz [1] "both the concept of rationality of a type of inference, and, generally, any concept of the correctness of this or that type reasoning requires relativization to the goal of that is taken into consideration. If no such relativization is carried out, then this goal is tacitly assumed. It is one of the most important tasks of methodology to find those aims which, in fact, determine an evaluation of the various mental operations and their product in the sciences, if it is to understand the sciences as cultural phenomena in the humanistic sense" (p. 249–250).

In most cases, both methods and goals must be hypothetical, since the scientist himself is seldom conscious of them. He learns his trade mostly by imitation. Therefore, when asked about either his purposes or methods, he usually responds in a metaphorical language, hardly satisfactory from the logical point of view.

There is no disparaging intention in this statement. Conscious pursuit of cognitive results by means of suitably adapted methods is neither necessary nor sufficient for obtaining those results. Excellent scientists have been known to express themselves in a hardly coherent way when they tried to describe their own work from meta-theoretic point of view. This led some authors to scepticism about the usefulness of methodological approach (cf. [8]). They advocate sociological and/or historical interpretation of science instead.

Another complicating factor is that scientific research is essentially creative and, as such, defies routinization. Possibly under Popper's influence, we are apt to concentrate on the creative aspects of inventing hypotheses. But designing experiments to test those hypotheses, or constructing necessary scientific apparatus, often requires no less imagination. Insofar as methods are patterns of action, to be used repeatedly, methods of discovery hardly exist unless, that is, we are prepared to stretch the (praxiological) meaning of the word 'method'.

The last remark suggests doubts as to the possibility of understanding science in terms of (even hypothetical) goals-means relation. Most that we can hope for is to approximate certain stages of scientific research by means of idealized descriptions. The risk of the description being inaccurate is not only unavoidable, but, which is much worse, non-measurable. An example of this is provided by the controversial issue whether it is more accurate to interpret beliefs of the scientist in terms of assertions or, rather, in terms of psychological probabilities. As the lottery paradox shows, the two interpretations cannot be reconciled. But we know of no sufficiently sharp criteria to decide which approach is a closer approximation to actual scientific practice.

With all these reservations made. I am going to discuss attempts to interpret rules governing the choice of hypotheses as means of a search for information. The discussion will be conducted in terms of the well known model, defined as follows.

Let $Q = \{h_1, \ldots h_N\}$ be an exhaustive set of pairwise exclusive hypotheses.

Let e be the empirical evidence, not necessarily fixed in advance, on the basis of which the choice of conclusion (see below) is to be made.

Let $p(h_k)$, $p(h_k/e)$ denote, respectively, the prior and the posterior probability of h_k (given e); $k = 1, \ldots, N$.

Let $A \subseteq 2^Q$ be (non-empty) set of the strongest admissible conclusions (cf. [2]) of the induction from e. Strictly speaking, for $a \in A$, the conclusion is a disjunction of all hypotheses belonging to a.

Let u_i be a real-valued function ('epistemic utility'), defined on $A \times Q$ and representing the value of the following situation: conclusion a is accepted as the strongest on evidence e, if hypothesis h_k is true. Obviously, for any such pair (a, h_k), the conclusion is correct if $h_k \in a$.

Let *cont* be a real-valued function on A, measuring semantics information of the conclusion a in A. The measure is negatively dependent on prior probability of a. In fact, *cont* is defined as the prior probability of the complement of a. It ought to be remarked that Levi [3] defines *cont* in terms of a measure which is not identical with prior probability; this measure is generated by the number of elements in a only.

So much for the concepts adopted. The set of conclusions A may be defined in several ways, subject to the obvious condition that it contains at least two elements and that the sum of its elements be Q. For instance, it may be a partition of Q. Extreme cases are, however, specially important:

$$A = 2^Q - \varnothing \text{ (} A \text{ unrestricted)}, \tag{1}$$

$$A = \{h_k\}_{k=1,\ldots,N} \text{ (} A \text{ maximally restricted)}. \tag{2}$$

Case (1) is adopted in Levi's analysis. Indeed, it is essential to his argument that truth cannot be the only aim of research. For, if we define epistemic utility as expressing precisely such aim, i.e. if we assign conventionally the value 1 to correct conclusion, and 0 to error:

$$u_1(a, h_k) = \begin{cases} 1 \text{ if } h_k \in a \\ 0 \text{ if } h_k \notin a, \end{cases}$$

then the best conclusion, under any evidence, is the maximal disjunction, i.e. Q itself, since its probability is always equal to 1.

This absurd consequence leads Levi to the introduction of information as additional goal of research. Under a plausible and rather weak condition (cf. [6]), the following function expresses the view that information is the goal of research:

$$u_2(a, h_k) = \begin{cases} \text{cont}(a) \text{ if } h_k \in a \\ \text{cont}(a) - 1 \text{ if } h_k \notin a. \end{cases}$$

If we define *cont*(a) as the prior probability of a's complements, then maximization of u_2 by a suitable choice of a is easily shown to be equivalent to the choice of such a for which the difference between its posterior and prior probability is maximal.

What Levi himself advocates is a compromise between the search for truth and the search for information. The nature of the compromise is determined by a constant q ($0 \leq q \leq 1$), expressing the willingness to risk. Resulting epistemic utility is a mixture of u_1 and u_2 (see [6] for details):

$$u_3 = (1 - q)u_1 + qu_2.$$

The adoption of u_3 as the numerical representation of the goal of research leads to maximization of the difference between the posterior probability of a and its prior probability weighed by the parameter q.

I am restating here the argument, since it represents the paradigmatic expression of the view that the choice of hypothesis in the light of evidence is an information-seeking process. The approach consists in introducing information into the definition of the goal of research. Which, in turn, is necessitated by the paradoxical consequences of the more traditional view that the only aim of science is the search for truth. If this were indeed the case (so the argument goes), science would never venture beyond the safe platitudes.

The plausibility of this approach is strengthened by the fact that the formal model above has a close analogue in actual scientific practice, viz. statistical estimation by means of confidence intervals. It has been noticed quite early (cf. [4], p. 255 ff.) that if the choice of the level of confidence were governed exclusively by the desire to have as reliable a conclusion as possible, then the only sensible decision would be to maximize this level, i.e. to make it equal to 1. This would fully protect the inference-maker from committing an error, but only at the price of making the conclusion tautological. The (average) length of confidence interval is essential in this connection. The smaller the interval, the more information it conveys about the magnitude to be estimated. The choice of confidence level appears, therefore, to be a compromise between maximization of reliability of the conclusion and maximization of information it conveys. The analogy with the epistemic utility function u_3 is obvious.

Plausible as it sounds, the argument is opened to serious objections. Against interval estimation, L.J. Savage, in his classical book [5], has pointed out that if accepting a conclusion is interpreted as acting upon it, then a definite value of the estimated parameter must be singled out, even though we known that error is inevitable (which happens when the set Q is not finite, as we had assumed, but continuous, as is the case *in* statistical estimation). To quote Savage (p. 261): "From the behavioralistic point of view, I maintain that point estimation fulfils as important function. On the other hand, I can cite no important behavioralistic interpretation of interval estimation".

Similar kind of reasoning applies if the problem is extended beyond the field of statistical inference proper, i.e. when the nature of hypotheses h_k is left unspecified, except for the postulate that they be *lawlike*, in the sense that they have explanatory or predictive content. Now, it is well known that a disjunction of lawlike hypotheses is not lawlike. It follows that such a disjunction is not satisfactory as a conclusion of inductive inference. In other words, the set A of possible conclusions cannot be unrestricted, i.e. defined as in (1).

Now, the assumption (1) was appealed to in the argument that the actual choice of hypothesis was governed by two conflicting desiderata: (i) that the conclusion be highly probable; (ii) that the conclusion be highly informative. Broadly speaking, this last property was measured (inversely) by the number of elements in the disjunctions accepted as the strongest conclusion. If the set A contains all disjunctions that can be formed out of elements of Q than the possible conclusions have unequal information which could be balanced against their posterior probability. With the rejection of (1), this line of reasoning is blocked.

Rejection of (1) is, of course, not equivalent to acceptance of (2). However, in what is going to follow I will assume that the set A of possible conclusions is identical with the initial set Q of (lawlike) hypotheses. This is certainly the simplest way to define A. Also, it seems realistic enough, since in most cases the problem of choice would not be considered as solved unless a single element of Q is selected on the basis of the evidence e.

If information (as the function '*cont*' quantified over sentences) is not introduced into the definition of epistemic utility, can we still make sense of the assertion that the purpose governing the choice of hypothesis is to obtain information? An attempt to give positive answer to this question could start from the concept of entropy of the set Q of hypotheses. Let us define, in the usual way, the absolute and conditional entropy of Q as

$$H(Q) = -\sum_k p(h_k)\log p(h_k), \tag{3}$$

$$H(Q/e) = -\sum_k p(h_k/e)\log p(h_k/e). \tag{4}$$

Then the difference

$$I(e, Q) = H(Q) - H(Q/e) \tag{5}$$

measures the degree to which uncertainty concerning Q has been lessened, thanks to the evidence e. Of course, this measure (due to Shannon) can actually be negative, in which case uncertainty as to which hypothesis is true would be increased, rather than decreased, by e. The standard reply to this objection consists in pointing out that (i) the average value of $I(e, Q)$ over all possible outcomes e of the experiment is non-negative; and (ii) under plausible assumptions, the conditional entropy $H(Q/e)$ tends to 0 when e is indefinitely increased.

The above defence of the 'entropy approach' is not quite convincing. When the quantity $H(Q/e)$ is actually big, it is small consolation to learn than 'in the limit' it could be made small; and that in the meantime we must choose a hypothesis from the set Q, no matter what our uncertainty is on the present evidence e. It seems, therefore, that some condition must be imposed on e, in the form of relation between e and the hypotheses it is supposed to test. Unless the condition is satisfied, the verdict 'no conclusion' will be a reasonable one. We are thus led to some kind of sequential rule of inference (cf. [7]).

Perhaps the most natural way to obtain such a rule is outlined by Jeffrey. I am going to quote, with inessential change in notation, the relevant fragments from his paper [2].

The generally accepted scientific hypotheses of the day – the 'laws' – are those serious competitors whose probabilities on the available evidence are orders of magnitude greater than those of the recognized alternatives. But the probabilities of such 'laws' must nevertheless be expected to be orders of magnitude less than those of their denials. (Remember that the denial of a scientific hypothesis is surely no scientific law – not because that denial is improbable but rather because it is not 'lawlike' e.g. it will essentially devoid of explanatory or predictive content (p. 113).

Empirical testing starts after a certain amount of would-be discovery has taken place: one or more hypotheses h_1, h_2, ... have been recognized as 'starters' in the race, i.e. as worth testing. These will be incompatible hypotheses whose denials are non-starters. The management is unable to cope with photo-finishes: a winner can be declared only if it is far and away the most probable of the starters after the empirical test has been made (p. 114).

If we are prepared to follow Jeffrey's suggestion we can express it formally as a rule of inference. Let B symbolize a constant satisfying $B \geq 1$. We adopt the maximally restricted set of possible conclusions: $A = Q$.

It will be convenient to shorten the posterior probability $p(h_k/e)$ to $p_k (k = 1, \ldots N)$. The B-rule. If

$$p_r \geq B \cdot p_s \quad \text{for all } s \neq r,$$

then h_r is accepted on the evidence e. If there is no hypothesis h_r satisfying the condition (B), then the evidence is considered insufficient for drawing a conclusion.

Clearly, the B-rule expresses Jeffrey's idea that the hypothesis accepted should be much more probable on the existing evidence than any of its competitors. Exactly how much, it is up to the inference-maker to decide by a suitable choice of B. If he adopts $B = 1$ he is back to the traditional rule of maximization of posterior probability. But this would be contrary to the philosophy outlined above. Therefore, the parameter B ought to be significantly higher than 1.

The B-rule has certain 'reasonable' properties. One of them is that the posterior probability of the hypothesis accepted according to B-rule cannot be lower than a certain threshold, depending both on B and on the number of hypotheses N. Suppose h_r is accepted on evidence e according to the B-rule. Then:

$$p_r \geq \frac{B}{B + N - 1}. \tag{6}$$

The inequality (6) is easily derived from condition (B), by summing all $s \neq r$. We thus see that the lower bound of p_r increases with B and decreases with the number N of hypotheses.

Also the posterior probability of any hypothesis h_s alternative to h_r, cannot exceed a value depending (negatively) on B:

$$p_r \leq \frac{1}{B + 1}. \tag{7}$$

This also accords with intuition.

The relation between the B-rule and the entropy of the set Q of the hypotheses is also not difficult to find. If we adopted the shortening

$$C = \frac{1}{B + N - 1}, \tag{8}$$

then the upper bound of entropy when h_r has been accepted on e according to the B-rule is determined by

$$H(Q/e) \leq -BC \log BC - (N - 1)C \log C. \tag{9}$$

This follows immediately from (6) and from the properties of entropy. By substituting this upper bound into

$$I(e, Q) = H(q) - H(Q/e) \geq 0, \tag{10}$$

and solving the equation for B, we can find the lowest value of the parameter B-such that the gain of information in the sense of $I(e, Q)$ would be ensured. The value of B would then depend on $H(Q)$, i.e. on the initial level of entropy of the set of hypotheses. This answers the objection against interpreting the inference procedure in terms of Shannon's measure of information. The objection was based on the fact that the empirical findings may actually increase the entropy of Q, thus making $I(e, Q)$ negative. The sequential B-rule has a built-in device against such a contingency. To be sure, there is a price to pay. Increasing the value of the B will, in general, lead to an extension of the empirical basis, which is not costless. I shall return to this point in my concluding remarks.

So much for the immediate properties of the B-rule. Can its use be justified in terms of epistemic utility? In other words, can it be shown that choice of hypothesis according to the B-rule is optimal relative to the aim of research as represented by a utility function u on $A \times Q$? The answer is yes; the class of such functions can be obtained in the following way.

Let $[u_{ij}]$ $N \times N$ be a (square) utility matrix satisfying minimal epistemic conditions:

$$u_{rr} = \text{Max}_i\, u_{ir} = \text{Max}_j\, u_{rj} r,\ i, j = 1, \ldots N, \tag{11}$$

so that a_r is the best choice if h_r is true. Therefore, a_r can be interpreted as accepting h_r.

Then, by definition of optimality, a_r is optimal relative to $[u_{ij}]$ on the evidence e if

$$\textstyle\sum_k u_{rk}\, p_k \geq \sum_k u_{sk}\, p_k \quad \text{for all } s. \tag{12}$$

In the above expression, p_k is, as before, a shorthand for $p(h_k/e)$. The condition (12) can be rewritten as

$$(u_{rr} - u_{sr})p_r \geq \textstyle\sum_{k \neq r} (u_{sk} - u_{rk})p_k \quad \text{for all } s. \tag{13}$$

It will be convenient to recall here the concept of loss (or regret) introduced in [5]. It has been defined as

$$w_{ij} =_{Df} \text{Max}_i\, u_{ij} - u_{ij}. \tag{14}$$

The loss w_{ij}, associated with accepting hypotheses h_i when in fact h_j is true, is thus equal to the difference between the maximal utility obtainable when h_j is true and the utility actually obtained. In view of (11), we can simplify (14) to

$$w_{ij} = u_{ij} - u_{ij}$$

In terms of thus defined loss, (13) can be rewritten successively as

$$(u_{rr} - u_{sr})p_r \geq \textstyle\sum_{k \neq r} [(u_{kk} - u_{rk}) - (u_{kk} - u_{sk})]p_k \quad \text{for all } s, \tag{16}$$

$$w_{sr}\, p_r \geq \textstyle\sum_{k \neq r} (w_{rk} - w_{sk})p_k \quad \text{for all } s; \tag{17}$$

$$p_r \geq \sum_{k \neq r} \frac{W_{rk} - W_{sk}}{W_{sk}}\, p_k \quad \text{for all } s. \tag{18}$$

The condition (12) for optimality of a_r is thus shown to be equivalent to (18). On the other hand, for h_r satisfying condition (B), we have

$$p_r \geq \sum_{k \neq r} \frac{B}{N-1} p_k. \tag{B^*}$$

The last formula is obtained by summing all inequalities in (B). From (18) and (B^*) it follows that a sufficient condition for (B) to entail optimality of a_r is

$$\frac{B}{N-1} \geq \frac{W_{rk} - W_{sk}}{W_{sk}} \quad \text{for all } s, \text{ all } k \neq r. \tag{19}$$

Since it is not known in advance which hypothesis will satisfy condition (B) relative to evidence e, condition (19) must hold *for all* r. The result can be summarized by the following *theorem* T.

Any $N \times N$ matrix of epistemic utility satisfying (19) for $r = 1, \ldots N$ has the property that if a hypothesis h_r is accepted on evidence e according to the rule of inference B, then h_r is optimal in the sense of the utility matrix, i.e. its acceptance maximizes the expected utility, relative to posterior probability distribution based on the evidence e.

Condition (19) deserves a comment, since its meaning is not immediately clear. The quantity to be kept under $B/(N-1)$ limit can be described as follows: it is the difference between the loss associated with unjustified acceptance of h_r and the loss associated with the acceptance of h_s, under any alternative hypothesis h_k (the last mentioned loss becomes 0 when $s = k$); the difference is normalized by means of loss associated with h^s under h_r.

The crucial quantity here is, of course, w_{rk}. If, for some k, accepting h_r involves a catastrophically high loss when h_k is true, then the probability of h_r must be much higher than that of h_k in order to offset the influence of this fact. This is what the condition (B) ensures, for B sufficiently high. Condition (19) imposes on the epistemic utility matrix a kind of uniformity, in the sense that 'epistemic losses' are of roughly the same order of magnitude.

Postulated uniformity depends on B: the lower the value of parameter B, the less variation of 'epistemic losses' is allowed. On the other hand, if B is high (hence, the hypothesis accepted is many times more probable than any of its competitors) then the accepted hypothesis is optimal even in such cases when the aim of research associates high 'penalty' with certain types of error.

Theorem T asserts that condition (19) is sufficient for optimality of h_r. It is, of course, not necessary. Suppose, for instance, that the epistemic utility matrix is the simplest, i.e. zero-one type; such utility function was defined as u_1 for the most general case of rectangular matrix. Then, if there are e.g. three hypotheses, B must be at least equal to 2 in order to satisfy condition (19). But obviously, a_r is optimal relative to u_1 if its posterior probability is maximum, which in turn means that h is accepted by the rule B, for $B = 1$. Therefore, in this case, condition (19) is stronger than necessary.

The above discussion can be summarized as follows. One way to express the view that science aims at increase of information is to introduce an information measure into the definition of epistemic utility. Conclusions are then compared in terms of their content, inversely related to their (posterior) probability. The aim of research is, therefore, a compromise between two conflicting values: information and truth [3]. This approach is based on the assumption that admissible conditions are disjunctions of 'simple' hypotheses: the larger the disjunction, the less content it has. A possible objection points to the fact that a disjunction of lawlike hypotheses is not, in general, lawlike.

An alternative way is to allow only simple hypotheses as possible conclusions and to interpret the search for information as maximization of posterior probability (by a suitable choice of conclusion). Entropy is thereby lowered (on average), hence the gain in information. Information is here conceived in the classical sense, due to Shannon.

An improved version of the last approach consists in replacing simple maximization of posterior probability by a stronger rule of choice. This leads to sequential procedure, since it may easily happen that, for a given evidence, no hypothesis satisfied the stronger requirement. Some properties of such rule of inference have been investigated in the present paper. In particular, it has been shown that the hypothesis selected according to this is optimal relative to epistemic utility function satisfying certain condition on uniformity.

It could be said that the case of the B-rule, the conflict between truth and information is replaced by another conflict, viz. that between truth and the economy of evidence. Since the resources for increasing evidence are not unlimited, we have to stop somewhere. The 'somewhere' finds expression in the parameter B.

REFERENCES

1. Ajdukiewicz K., "The Problem of Rationality of Non-Deductive Types of Inference", [in:] K. Ajdukiewicz, *The Scientific World – Perspective and Other Essays*, ed. by J. Giedymin, D. Reidel Publ. Co., Dordrecht 1978.
2. R. Jeffrey, "Critique of the Popper's Program", *Synthèse*, 1975, 30.
3. Levi I., *Gambling with Truth. An Essay on Induction and the Aims of Science*, A. Knopf, New York 1967.
4. Neyman J., *Lectures and Conferences on Mathematical Statistics and Probability*, Graduate School of U.S. Department of Agriculture, Washington 1952.
5. Savage L.J., *The Foundations of Statistics*, J. Wiley, New York 1954.
6. Szaniawski K., "Types of Information and Their Role in the Methodology of Science", [in:] M. Przełęcki, K. Szaniawski, R. Wójcicki (eds.), *Formal Methods in the Methodology of Empirical Sciences*, Ossolineum, Wrocław 1976 [see this volume, pp. 167–176.].
7. – "On Sequential Inference", [in:] R.J. Bogdan (ed.), *Local Induction*, D. Reidel Publ. Co., Dordrecht 1976. [see this volume, pp. 96–105.]
8. Ziman J., *Public Knowledge*, Cambridge University Press, 1968.

7

INFERENCE OR BEHAVIOUR?*

> The view has, however, really been advanced (Neyman, 1938) that Inductive Reasoning does not exist, but only Inductive Behaviour.
>
> (R.A. Fisher, 1956)
>
> It will appear that the term 'inductive reasoning' is a misnomer, contributing to the confusion regarding the nature of scientific research, and that a better term would be something like 'inductive behaviour'.
>
> (J. Neyman, 1957)

The title of this paper refers to a controversy concerning the logical status of the so-called statistical procedures, such as verification of hypotheses and estimation of parameters. The conflicting views are perhaps best represented in the writing of Sir Ronald Fisher and Jerzy Neyman. The problem under discussion is do the statistical procedures deserve the status of inferences or not? Neyman's opinion is that they are not inferences (or reasonings) at all and are better called inductive behaviour: according to him there is no such thing as inductive inference.[1] Fisher asserts the contrary.

In attempting to form an opinion on the point at issue one could, of course, begin by trying to answer the question: do the protagonists mean the same by the expression 'inference'? This is essentially a problem of interpretation, as a definition of the word 'inference' is not – so far as I know – to be found in the relevant texts. My guess is that the writers in question do not mean the same by this crucial word. If this guess is correct, the supposedly conflicting views would appear to be non-contradictory and the controversy we are concerned with would have arisen out of a rather trivial equivocation.

It seems, however, that to reduce this controversy to a purely verbal misunderstanding would be altogether inadequate. I suggest that different views concerning the justification of statistical procedures can be discerned behind the contentions that statistical procedures are (resp. are not) inferences. And the problem, how statistical procedures are to be justified, is certainly far from being spurious: it is not even devoid of some practical importance, as its different solutions lead, in general, to the adoption of different procedures.

The easiest way to show that this is the real point of the Neyman-Fisher controversy is perhaps by analyzing the argument supporting one of the two views. I have chosen the Neyman's view because it is much more clear and, what is more, it represents a unified theory of statistics, i.e. the so-called theory of decision functions.

The basic notions underlying this theory are quite easy to outline; which does not mean, of course, that the theory itself is mathematically unsophisticated. In what is going to follow several simplifications will have to be made.

Consider a population characterized by a parameter the actual value of which is not known. For instance, the population of the inhabitants of Cambridge and the proportion of cigarette smokers among them. Let this parameter be ω and the set of all its possible values Ω. Suppose further that we have to choose between several alternative actions (decisions) a; let the set of all such actions be A. For instance, we may have to decide what the weekly supply of cigarettes to Cambridge is going to be.

Each such decision in conjunction with the true value of the parameter ω leads to a situation (ω, a). Suppose that the disutility or loss associated with such situation can be measured and is expressed by a function $L(\omega, a)$, called loss function.[2] Of course, by a suitable choice of α we want to make the value of this function as small as possible.

The statistical problem arises if we want to base the choice of a on some partial information concerning the true value of ω; the information being provided by the outcome of an experiment designed for the purpose.[3] In our example the experiment might consist in choosing at random a certain number of inhabitants of Cambridge and ascertaining how many of them are cigarette smokers. The result x of such an experiment depends on ω: let us designate the probability distribution function of x by $p(x/\omega)$. Of course, $\sum_{x \in X} p(x/\omega) = 1$, where X is the set of all possible outcomes of the experiment.

Now the decisive step is to associate with each possible outcome of the experiment some particular decision a. This is equivalent to adopting some function $d(x) = a$, defined for all x's and taking values out of the set A; d is called decision function. But there is a considerable number of such functions and the statistical problem is, which one of them to choose. This problem of choice is the statistical problem, according to the theory of decision functions.

Its solution must depends on the expected value of loss when the true value of the parameter is ω and the decision function adopted is d. Expected value of loss is called the risk function, $r(\omega, d)$, and can be calculated as

$$r(\omega, d) = \sum_{x \in X} [L(\omega, d(x)) \cdot p(x/\omega)]$$

The risk is thus a function of both ω and d, the value of ω being unknown.[4]

Several possible criteria of choice in such a situation have been suggested. I am not going to discuss in the present paper their respective merits. Anyhow, every possible criterion of choice of the argument d of the risk function r provides a solution to the statistical problem – as defined in the theory of decision functions.

It is easily seen that the essential characteristics of the approach outlines above are the following.

1. A statistical procedure (under the assumption that the experiment is completely defined in advance) consist in adopting a decision according to the outcome of the experiment. It can thus, from a formal point of view, be identified with a decision function.
2. The choice of decision function is governed by the tendency to minimize the expected loss, $r(\omega, d)$. This, however, depends also on ω the value of which is not known. The choice of d is not, therefore, determined in a unique way by the data of the problem.
3. Whichever criterion of choice is adopted, the actual choice of decision function always depends on loss function.

Now, it is said by some mathematicians, notably Neyman, that if a statistical procedure is understood in this way, it is not an inference. Apparently, there are, broadly speaking, two reasons for this assertion.

I. Every inference ought to lead to the acceptance of a statement or hypothesis, while the procedure described in the theory of decision functions leads to a choice of action or practical decision. Indeed, it might be said that according to Wald's theory the problem itself seems to be: 'how to minimize the loss function?', rather than: 'how to derive a true conclusion?'.

This argument does not seem to be very convincing. In the first place, it ought to be noted that in some cases the decisions can be interpreted directly as decisions to accept some hypotheses concerning the true value of the parameter ω. For instance, if the set A consists of two elements a_1, $a_2 a_1$ may be interpreted as the decision to accept the hypothesis that $\omega \leq \omega_0$, while a_2 – that $\omega \geq \omega_0$.

But even if the a's represent genuine practical decisions, it is possible to interpret the choice of an a as an implicit acceptance of hypothesis that $\omega \in \Omega_a$, where $\Omega_a \subset \Omega$ and is defined in the following way:

$$\Omega_a = df(\omega)\{(\omega \in \Omega) \wedge [L(\omega, a) = \min_{\omega'} L(\omega', a)]\}.$$

Ω_a is thus the set of all those ω's which minimize the loss function under the choice of the action a.[5] Every decision function would then lead, directly or indirectly, to the acceptance of a hypothesis concerning the true value of the parameter ω, whatever the word 'acceptance' may mean.

II. The second reason for denying the statistical procedures the name of (inductive) inference is perhaps more complex. Broadly speaking, it is argued that in a statistical procedure the final acceptance of a hypothesis concerning the true value of a parameter is largely a matter of arbitrary choice, an 'act of will' (Neyman). And this appears to be incompatible with the meaning of the world 'inference'.

It would seem, therefore, necessary to inquire: just how arbitrary is the choice of so-called conclusion in the so-called statistical inference? It may perhaps be said without oversimplifying the issue (still under the assumption that the exact character of the experiment is decided upon in advance) that three arbitrary elements can be detected:

1. the initial assumptions,
2. the loss function,
3. the criterion of choice of decision function.

1. By 'initial assumptions' I mean assumptions concerning the actual form of the probability distribution function $p(x/\omega)$. Is it, for instance, binomial or Poissonian? What are the values of other parameters (if any) of the distribution? There may be also some restrictions on the set Ω of the possible values of the parameter in question. In general, initial assumptions specify the class of probability distributions to which the actual probability distribution belongs. Such assumptions summarize our 'background knowledge' (as Popper calls it), more or less firmly established. Without any such background knowledge, however hypothetical, statistical procedure is seldom possible.

Now, it is argued that the adoption of initial assumptions depends to a great extent upon the scientist's discernment and is anyhow – in a large number of cases – a matter of more or less arbitrary choice, which way be later on justified or not. It is a sort of guess-work, guided perhaps by some heuristic rules.

It would be perhaps difficult for an 'inductionist' to answer this challenge in a fully satisfactory way without making himself explicit on the 'eternal' problem of the justification of induction. However, one possible line of defence may be indicated here. It can, namely, be argued that the possible arbitrariness of initial assumptions is irrelevant to the issue under discussion. Be the initial assumptions arbitrary to some degree or not, the problem is – just like in the case of deductive inference – what is the relation between the premisses (including such assumptions) and the conclusion. Is this relation of such a nature that it leaves large (how large?) scope for the choice of conclusion by the scientist? In other words: no matter what have been the reasons for adopting such and such particular premisses, in what way does the choice of conclusion depend on them?

Opponents might argue again that this is not what is called 'induction' in the strict sense of the word. That we have no right to include the initial assumptions among the premisses; that the only premisses we start from in the induction proper are the protocols of what has been actually observed, that is: an answer to the question, which one of all the possible outcomes of the experiment actually occurred.

But to this a possible reply is that the situation in which nothing but direct experimental evidence is available, is highly artificial. For a philosopher who wants to answer the question, how do we arrive at any general empirical knowledge, induction in this sense may be of primary importance. For someone, however, who wants to analyse inductive procedures as they are actually performed – that is, starting from some, more or less hypothetical, knowledge of the population in question – the relation between such premisses and the conclusion is what really matters.

If this is granted, the problem of statistical procedure being an inference or not may be reformulated as follows. Consider the rule based on a decision function $d(x) = h$; the rule is: if the outcome of the experiment is x, then accept the hypothesis h. Is this rule to be called a rule of inference?

The answer provided by the 'anti-inductionist' school would still be: no. The rule itself – so it is asserted – is chosen in an arbitrary way; therefore, it is a matter of arbitrary decision which hypothesis is eventually accepted as conclusion.

We shall have to consider in some detail the grounds for this assertion.

2. It is argued that the choice of conclusion depends to a considerable degree upon the loss function. But the choice of the loss function itself seems to be arbitrary: it may be dictated by some considerations other than the tendency to derive the true conclusion.[6] Therefore the choice of conclusion is subject to a somewhat arbitrary decision.

The question arises, is it always the case? The situation is somewhat different for (a) the (slightly generalized) verification of hypotheses, and (b) estimation of parameters.[7]

(a) Suppose that the set Ω of all possible values of ω is defined as: $a \leq \omega \leq b$. Let us divide this set into a finite number of intervals: $\Omega_1, \ldots \Omega_n$; they will represent the hypotheses considered, i.e. the ith hypothesis asserts that $\omega \in \Omega_i$.

Then for every value of the parameter ω we can calculate the probability of accepting the hypothesis Ω_i by using decision function d, say $P(\Omega_i/d, \omega)$.

If we introduce

$$X_{d\Omega i} = df(\hat{x})[d(x) = \Omega_i],$$

then

$$P(\Omega_i/d, \omega) = \sum_{x \in X_{d\Omega i}} p(x/\omega).$$

But for every ω there is exactly one such Ω_i that $\omega \in \Omega_i$. Therefore, it is possible to compute the probability, say $P(\text{ver}\langle d\rangle/\omega)$, of accepting the true conclusion under the assumption that the true value of the parameter is ω. Of course, we want this probability to be as large as possible. As this probability depends upon two factors, ω and d, we are again facing a decision problem under uncertainty with the loss function $1 - P(\text{ver}\langle d\rangle/\omega)$.

So far we have not assumed explicitly any loss function. But from a formal point of view our procedure is equivalent to adopting the loss function which with a right conclusion associates the loss 0 and with a wrong one, the loss 1. There is, therefore, nothing arbitrary in adopting such loss function (called by Wald 'simple'). Indeed, the notion of loss function can in this case be dispensed with. What we are doing here, is only computing the probability of error, given a decision function and the true value of the parameter.

Any other loss function is introduced only if we want to avoid some kinds of error more than others (and this may by due not necessarily to practical reasons, especially if the number n of intervals is large). Thus it may happen that if the hypothesis Ω_i is true, the statistician prefers the error of accepting Ω_j to that of accepting Ω_k. He then must adopt such loss function L that L (Ω_l, accepting Ω_j) $<$ L(Ω_l, accepting Ω_k). The loss function is then used as a measure for different kinds of error; perhaps it would be more appropriate to call it, as Wald did, weight function rather. Of course, in such a case there is some scope for expressing personal preferences and the element of arbitrariness does then enter into the picture. But there is no necessity for adopting any other than simple loss function.

(b) Assume again that the set Ω of possible values of ω is: $a \leq \omega \leq b$; but this time it is identical with the set of hypotheses, i.e. each hypothesis is of the type $\omega = \omega_0$, where $\omega_0 \in \Omega$.

Then the probability of accepting ω when actually ω is the true value of the parameter is (under very weak assumption) equal to 0 for all decision functions d.[8] This means that there are no chances at all of arriving at a true conclusion. It is therefore necessary to consider it a success if the accepted value of the parameter is sufficiently near its true value. Thus the purpose of statistical procedure becomes that of minimizing the distance between the accepted value of the parameter, say $\hat{\omega}$, and its true value ω; this is equivalent to adopting as the loss function: $|\omega - \hat{\omega}|$, or the Gaussian $(\omega - \omega)$,^{2} or some other monotonically increasing function of the distance. In this way the arbitrariness in the choice of loss function becomes drastically limited, if no other considerations (such as, for instance, monetary losses and gains from a contemplated action) are additionally introduced.

To sum up this part of the discussion. It appears that loss function is not always necessary. True, that in some (rather important) cases it is; and this, incidentally, undermines Fisher's assertion that loss function is something the statistician can always do without. But the actual choice of loss function is not necessarily dictated by the prospect of material losses and gains. And if the purpose of statistical procedure is to minimize, in a sense, the size of error (the case of point-estimation), then the arbitrariness in the choice of loss function can be seriously reduced.

3. We come at last to the criterion of choice of decision function. It must be said that this is the only arbitrary step that cannot be avoided. The point is that as the actual value ω in $r(\omega, d)$ is not known, the postulate of minimizing the risk function by a suitable choice if d does not lead to a unique solution.

Of course, some conditions can be formulated, to be satisfied by any criterion of choice of d. But they in turn are to some extent arbitrary.

The only condition unassailable in view of the postulate that the risk function be minimized, is the following one. Suppose that for all ω: $r(\omega, d_1) \leq r(\omega, d_2)$ and for some ω: $r(\omega, d_1) < r(\omega, d_2)$. Then d_2 is said to be dominated by d_1, and it is clear that d_2 should not be adopted in preference to d_1. Thus an important problem is that of the existence of the class of decision functions not dominated by some others. A large part of statistical literature is devoted to the discussion of this problem. But even if such a class of decision functions exists, the question still remains, which one of them is to be adopted.

Every answer to this question, i.e. every criterion of choice of decision function, must be more or less arbitrary. It appears, therefore, that if we are to interpret decision rules as rules of inference, the choice of a rule of inference is indeed arbitrary within certain limits. This is the only unavoidable arbitrariness in statistical procedure if the expression 'statistical procedure' is taken in a somewhat restricted sense, that is not including the choice of initial assumptions.

Is this arbitrariness of such a nature that it violates the meaning of the word 'inference'? This depends upon what is meant by 'inference' and is perhaps not a problem of particular importance. As long as we realize what we are doing, it does not matter greatly how we call our activity.

The real point at issue is thus not a terminological one. When protesting against the use of the expression 'inductive inference', Neyman in fact defends the view that statistical procedures can best be described and, in a sense, justified within the conceptual framework of the theory of decision functions. And this is precisely what R. Fisher objects to, while maintaining that statistical procedure are to be treated as inferences. It hardly needs stressing that this issue is of considerable importance, both from the theoretical and the practical point of view. It is, therefore, to be regretted that this very real issue has been often obscured by an unfortunate wording, creating the impression that the difference of opinions is of a verbal nature.

I do not intend to discuss here the arguments for or against the theory of decision functions as adequate basis of statistics. This is a rather complex problem and it deserves of detailed analysis. An outline of the controversy can be found in *Games and Decisions* by Luce and Raiffa on pp. 318–324.

NOTES

* This paper has been read in May 1958, at a meeting of the Moral Sciences Club, Cambridge (England).

[1] This view is shared by some logicians, notably K. Popper.

[2] Actually, the assumption that disutility is measurable is not such a strong one as it might seem at a first glance. Within the limits defined by certain postulates (cf. J. von Neumann and O. Morgenstern: *Theory of Games and Economic Behaviour*, Princeton 1947, p. 26–29; R.D. Luce and H. Raiffa: *Games and Decisions*, New York 1957, p. 23–32) the choice of loss function does not matter. Thus, for instance, if for a given problem $L(\omega, a)$ is the loss function, then $a \cdot L + \beta$, where $a > 0$ and β is any number, could do just as well.

[3] The design of the experiment, e.g. its size (the number of trials), is a matter to be decided upon and it might be influenced by such factors as the cost of experimenting and the results obtained in successive phases. We shall, however, simplify the situation by assuming that the experiment to be performed is well defined in advance.

[4] The problem of choice of decision function, given the risk function the value of which is to be minimized in some sense, is a special case of the so-called decision problem under uncertainty. It is assumed that no a priori information concerning the value of ω (such as, e.g., a probability distribution over Ω) is available. Cf. K. Szaniawski: "Some Remarks Concerning the Decision Problem under Uncertainty", *Studia Logica IX*, 1960.

[5] Of course, the family of sets Ω_a is not, in general a partition of Ω: the sets Ω need not be disjoint and their sum may be a proper subset of Ω.

[6] It should be noticed that the opponents of Neyman's view make use of the same argument in order to show that the theory of decision functions is inadmissible. Thus R. Fisher writes: 'It is important that the scientific worker introduces no cost functions for faulty decisions, as it is reasonable and often necessary to do with an Acceptance Procedure. To do so would imply that the purposes to which new knowledge was to be put were known and capable of evaluation. If, however, scientific findings are communicated for the enlightment of other free minds, they may be put sooner or later to the service of a number of purposes, of which we know nothing. As workers in Science we aim, in fact, at methods of inference which shall be equally convincing to all freely reasoning minds, entirely independently of any intentions that might be furthered by utilizing the knowledge inferred' (*Statistical Methods and Scientific Inference*, Edinburgh 1956, p. 102–103).

[7] For the sake of simplicity only point-estimation will be treated here.

[8] Cf. K. Szaniawski, "On Some Basic Paterns of Statistical Inference", *Studia Logica XI*, 1961, p. 83 [see this volume, pp. 70–79.].

8
A NOTE ON CONFIRMATION OF STATISTICAL HYPOTHESES

1. The purpose of this paper is to discuss some aspects of confirmation as defined by Professor Popper in his stimulating series of articles published in '*The British Journal for the Philosophy of Science*' in the years 1954–1958.[1] The main problem under investigation will be: is the suggested measure suitable for accessing the degree to which a statistical hypothesis is confirmed by an experimentally verifiable test-statement? Popper asserts that it is; in fact, his Third Note (V) is almost entirely devoted to the task of justifying this assertion. It seems, however, that several doubts may be raised concerning the validity of Popper's argument.

2. As possible explication of the expression 'the degree to which a statement e confirms a statement h' Popper suggests

$$E(h, e) = \frac{P(e, h) - P(e)}{P(e, h) + P(e)}, \tag{1}$$

or (better still):

$$C(h, e) = \frac{P(e, h) - P(e)}{P(e, h) - P(e \cdot h) + P(e)}. \tag{2}$$

The above definitions are appraised in the light of a set of postulates for the concept of confirmation, which I am not going to quote here. (See I, p. 147–8.) It appears that C fulfills all of the desiderata while E – only some of them (not the least important ones, though).

The symbol $P(x)$ is to be interpreted as the *logical* probability of x; $P(x, y)$ means: the conditional, (logical) probability of x given y, and is defined in the usual way as

$$P(x, y) = \frac{P(x \cdot y)}{P(y)}. \tag{3}$$

The concept of logical probability is characterized axiomatically and some possible ways of doing this are indicated by Popper in his other works. (See, e.g. VI.)

3. So far, the position seems to be quite clear, with the possible exception of the rather doubtful notion of logical probability. If I am not mistaken, no general way of calculating $P(x)$ for each given x is known. If we rule out the case when x is a sentence asserting that a random variable takes on some given value (because then the logical probability of x is practically undistinguishable from the statistical one), the problem of determining the value of $P(x)$ seems to be solved only for extremely simple, not to say trivial, cases. And if we do not know how to assess the values of $P(e)$, $P(e, h)$ etc. how are we to calculate $E(h, e)$ or $C(h, e)$?

This is not, however, the main issue I wish to raise. I mention my doubts concerning the general applicability of the concept 'logical probability' only in order to stress the importance of the problem Popper has taken up in his Third Note: is it possible to measure the degree to which a statistical hypothesis is confirmed by experimental evidence relevant to it? Obviously this is the case where we (more or less) know how to calculate probabilities. The confirmation of statistical hypotheses by experimental evidence is, therefore, the touchstone of every definition of confirmation in terms of probabilities.[2]

3. In what is going to follow I shall assume throughout that the symbols like $P(e)$ or $P(e, h)$ are interpreted statistically in the usual way, i.e. that they are to be calculated in terms of relative frequencies. It might be objected that Popper expressly forbids such an interpretation, when he writes: ". . . the term 'logical probability' is better used for the 'logical interpretation' of both $P(x)$ and $P(x, y)$, as opposed to their 'statistical interpretation' which may be ignored here", I, p. 143). But in the first place, if we are to apply the concept of confirmation to statistical hypotheses then it is not clear in what way would the 'logical probabilities' differ from 'statistical' ones; and how would their numerical value be determined if not in the usual way, i.e. by relative frequencies? Secondly, the actual calculations to be found in V seems to show that this is what Popper in fact does when he ascribes a value r to the probability of a coin's falling heads.[3] The fact that he consistently speaks of statements rather than of events does not present a serious difficulty as there is a one-one correspondence between the event of a coin's falling heads and the sentence asserting precisely this.

4. I am now going to examine the problem of applying Popper's definitions (1) and (2) to a simple procedure of verification. In order to this I shall use the example discussed at some length in V.

Suppose that we want to verify a hypothesis h stating that the probability p of certain penny's falling heads in a toss randomized by spinning is 1/2 (in general: r). In order to verify h we are going to perform an experiment consisting of n independent trials. By the (statistically interpreted) outcome x of the experiment we shall mean the proportion of heads in this series of trials. Obviously, the set X of all possible outcomes consists of $n + 1$ elements: $\{0/n, 1/n, \dots n/n\}$[4].

The problem now is to choose such a subset e' of X (symmetrical in respect to r) that if x falls into e', we shall say the experiment has confirmed (corroborated) our hypothesis. Let the subset consist of all x such that $r-\delta \leq x \leq r+\delta$; and the sentence stating that the actual outcome x belongs to e' we shall designate by e; there is, of course, a one-one correspondence between the subset e' and the statement e that x belongs to e'.

Now, according to Popper, we can calculate the degree to which h is confirmed by e; this will be given by the value of E or, if possible, C. I am afraid, however, that some serious misunderstandings are implied in the procedure of applying the formulae for E and C to the case under consideration.

5. There is no doubt that the symbol $P(e, h)$ is interpreted by Popper as the conditional probability of e given h. This is clear, e.g., from the following passage in V: 'It should be noticed that the absolute logical probability of a universal law h – that is $P(h)$ – will be in general zero, in an infinite universe. For this reason, $P(e, h)$ – that is, the likelihood of h – will become indefinite, in most systems of probability, since in most systems $P(e, h)$ is defined as $P(e \cdot h)/P(h) = 0/0$. We therefore need a formal calculus of probability which yields definite values for $P(e, h)$ even $P(h) = 0$, and which will always and unambiguously yield $P(e, h) = 1$ whenever e follows (or 'almost follows') from h". (V, p. 301).

Thus, according to Popper, $P(e, h)$ is defined as:

$$P(e, h) = \frac{P(e \cdot h)}{P(h)}, \tag{4}$$

where e is the report on the outcome of the experiment stating that $x \; c \; e'$, and h is the hypothesis stating that p (i.e. the probability of given penny's falling heads) equals some real number r. But surely if we are to speak of $P(h) = P(p = r)$, we shall have to make the assumption that p is a *random variable* assuming certain values with certain probabilities; $P(p = r)$ is precisely the probability of this random variable's assuming the value r (otherwise: see 7). Speaking in terms of our model this is equivalent to the assumption that the coin in question – let us call it z – had been selected, prior to the experiment, from a (possibly infinite) set Z of coins, each of them characterized by the probability of its falling heads. And $P(p = r)$ is the probability of such a coin being selected from Z which has the probability r of falling heads (if the choice of z was made at random $P(p = r)$ is, of course, the proportion in Z of coins having the probability r).

Now, there was absolutely nothing in the description of our problem to suggest the existence of such a set Z. We could, of course, postulate the existence and uniqueness of Z. We could also make some assumptions concerning the probability distribution of p. Such assumptions, however, seem to me – in general – quite arbitrary and unwarranted by the data of the problem.

I certainly do not want to say that the two-stage lottery model is never applicable: in some situations it is. In general, however, we have no grounds for implicitly assuming that the population we are dealing with has itself been chosen at random from a hyper-population.

The difficulty in question arose from our having interpreted $P(e, h)$ as the conditional probability of e given h. In view of what has been said above it seems to me better to drop this interpretation.

Speaking more generally: there is a big difference between verifying a statistical hypothesis when the parameter in question is random variable itself (incidentally: then and only then the Bayes theorem can, in principle, be applied) and the case when it is not. In the second case (which is perhaps more important as it seems much more frequent) the probability of an outcome of the experiment given the hypothetical value of the parameter is certainly not a conditional probability: it is simply the probability of the outcome, calculated under the assumption that the hypothesis concerning the distribution is true.[5]

Therefore, in the case of our hypothesis h which states that $p = r$ (and bearing in mind that the trials the experiment consists of are independent) $P(e, h)$ can be defined in exactly one way, i.e.:

$$P(e, h) = \sum_{n(r-b) \leq k \leq n(r-\delta)} \binom{n}{k} r^k (1-r)^{n-k}. \tag{5}$$

This interpretation of $P(e, h)$ I shall adopt below. It ought to be remarked that Popper himself calculates the actual value of $P(e, h)$ according to (5); he writes: '. . . in this case, it is adequate to assume a combinatoric, i.e. Bernoullian, rather than a Laplacean distribution' (V, p. 299).

6. Another misunderstanding seems to me to be worth clearing in this connection. Popper repeatedly calls $P(e, h)$ 'Fisher's likelihood of h' (See V, p. 298, 300-301). And he considers the expression

$$P(e, h) - P(e) \tag{6}$$

to be a generalization of Fisher's likelihood, more suitable (in general) as a measure of degree of corroboration. Now, I do not know if Fisher intended his likelihood function to serve as such a measure. But I am quite certain that he intended it to be a tool in solving the problem of point-estimation, which is – in logical terms – the problem of choosing a (so-called simple) hypothesis for a given e. In consequence, his likelihood function $P(e, h)$ of h, for a *fixed* e (e itself is a function of the actual outcomes of trials the experiment consists of). It is easy to see that this is something quite different from the problem of verifying a *given* h by a suitably test-selected test-statement e.

It might be added that from the point of view of choosing some h (from the set of admissible hypotheses) the expression (6) is no improvement on Fisher's likelihood, in whatever way $P(e)$ is defined, because it selects precisely the same h as $P(e, h)$ does. Fisher's rule of choice (the principle of maximum likelihood) is to accept this h which maximizes the likelihood function. But maximizing $P(e, h)$ or $P(e, h) - P(e)$ with respect to h leads to the same value of h if we bear in mind that throughout the process e remains fixed.

7. If p is not a random variable (and I shall assume from now on that it is not) we can still, of course, speak of a (statistical) probability of h; meaning by $P(h)$ the frequency of our coin's z having the probability r of falling heads. But

then $P(h)$ is either 0 or 1; and we go to the trouble of verifying h precisely because of the fact that we do not know which is the case. If this is so, then it is not clear how are we to interpret the expression $P(e \cdot h)$ in the denominator of (2). Being in doubt on that point I shall restrict my considerations to the function E.

8. In defining a measure for 'the degree of confirmation of h by e'the expression (6) plays obviously a crucial role. In Popper's words, $E(h, e)$ is but a way of 'normalizing' the expression (6).[6] Having discussed the problem of interpreting $P(e, h)$ we have, therefore, to turn now to the expression $P(e)$, 'the absolute probability of e'.

Now, it is not immediately obvious what 'the absolute probability of e' might mean. Popper's suggestion is that we ought to calculate $P(e)$ by ascribing equal probabilities to the $n + 1$ possible outcomes of the experiment. (Let me remind that by an outcome x I mean the relative frequency of successes in the series of trials; or, what Popper calls a statistical report about the sample. Thus there are $n + 1$ different outcomes possible). Leaving aside the question that it is difficult to see any justification for making such an assumption (i.e. for postulating the equidistribution on X), it ought to be noted that this assumption directly contradicts an important datum of the problem, i.e.: that all the trials in the experiment are *independent*. (This is rather obvious. If the trials are independent, each of them yielding positive result with the same constant probability r, then in the hypothetical 'long' series of experiments – consisting, each of them, of n such trials – the relative frequency of experiments yielding, say, k successes cannot br equal to $1/(n + 1)$ for every k).

Let me be more explicit on that point. We are perfectly entitled to consider any hypothesis alternative to h (like, for instance, that the probability in question is not r but some other value s) *except* such a hypothesis which violates the assumptions of the problem itself, e.g. the description of the experiment. Once we have assumed that the trials the experiment consists of are independent we shall have to respect this assumption throughout our discussion of the bearing of its outcome on the hypothesis h. It is for this reason that I would not be prepared to consider the assumption of equidistribution on X admissible.

But if this is the case, how are we to interpret the expression $P(e)$? I am leaving the question open.

9. Despite what I have just said I will, for the moment, assume the validity of calculating $P(e)$ from equidistribution on X and I shall try to see its consequences.

According to Popper we shall say that

e supports h if and only if $P(e, h) - P(e) > 0$;
e undermines h if and only if $P(e, h) - P(e) < 0$.

And the confirmation of h by e is the better, the greater the value of $P(e, h) - P(e)$.

In above expressions e is a test-statement of h, i.e. such a statement of the form $r - \delta \leq x \leq r + \delta$ that if it is experimentally found to be false we are forced to reject h. I take it, therefore, that every e (or, equivalently, every δ) determines what is called in statistical parlance a 'critical region' or 'region of rejection' of the hypothesis h. It determines also a 'region of acceptance', i.e. such a subset of the sample space X that if the actual outcome belongs to it, we may – always tentatively – accept h. (Cf. I, p. 146). We denote this subset by e'.

But how, it might be asked, the value of δ is to be chosen (for a fixed and presumably high n)? Is it a matter for a (more or less) arbitrary decision? If I understand Popper's argument correctly, his suggestion is that we should choose the value of δ for which the corresponding test-statement e confirms h to the highest degree attainable under the circumstances.[7] In other words, we should choose such e (hence: such δ) which maximizes the expression $P(e, h) - P(e)$. Alternatively, we might want to maximize a non-decreasing function of this expression, such as $E(h, e)$.

We shall now have to consider briefly the determination of the acceptance region e' and perhaps to examine some of its properties.

10. Suppose for the moment that n is fixed (we may as well assume that it is large) and consider the choice of e Popper recommends, i.e. such that (a) $E(h, e)$ or (b) $P(e, h) - P(e)$ be maximum.

Let us remark in the first place that if $P(e, h)$ is defined as in (5) and $P(e)$ is calculated as $1/(n + 1) \cdot l$ where l is the number of x's belonging to e' (that is, fulfilling the inequality $r - \delta \leq x \leq r + \delta$), then the expression $P(e, h) - P(e)$ is non-negative for every e; this means that no such test-statement of the form $x \in e'$ can be constructed which would undermine h in Popper's sense of the word 'undermining'. The smallest possible value of $P(e, h) - P(e)$, and also of $E(h, e)$, is zero; this is obtained for $e' = X$ (in statistical terms: when the critical region is an empty set).

Secondly, there is a significant difference between maximizing (a) and (b). We shall consider them in turn.

(a) It is not difficult to prove that with δ decreasing $E(h, e)$ increases and reaches its highest value – which for large n will be nearly 1 – for the smallest δ possible (by 'possible' I mean such δ that the e' is not empty). Thus, for instance, if n is even, say equal to $2k$, and $r = 1/2$, then $E(h, e)$ reaches its maximum for e' containing only one element, i.e. $k/2k = 1/2$.

Now, such "ideal evidence" (to use Popper's words) is rather difficult to have, especially for large n. If we choose, therefore, as our test-statement such e which maximizes $E(h, e)$, we shall be forced to reject h in an overwhelming proportion of cases, practically always for large n, when $P(x, h)$ approaches normal distribution. This does not seem reasonable, as any value of x 'close enough' to r (but not necessarily the closest) would be generally considered a sufficient confirmation for h, n being fixed and large.

Obviously such a choice of e would be contrary to the rule Professor Popper adheres to i.e. that $P(e, h)$ be high, e 'almost following' from h. And the rule is well justified: if it is violated, it is highly probable that h will be rejected when it is in fact true.

We are thus left with the alternative of maximizing $P(e, h) - P(e)$.

(b) If we determine e by the maximization of $P(e, h) - P(e)$ we do obtain a test-statement e to which the above objection does not apply. Thus, for instance if $n = 30$ and $r = 1/2$, δ ought to be 4/30 which gives the test-statement e: $11/30 \leq x \leq 19/30$. Are we now to consider this test-statement the best one for verifying (or rather: attempting to falsify) h when n is equal to 30? In what sense it is 'best' one?

Possible reply would be that such e confirms h to the highest degree attainable under the circumstances, the concept of 'confirmation' itself being justified by the set of desiderata in I. But even if we leave aside the objection concerning 'conditional probabilities' and the way of calculating $P(e)$, the desiderata in question apply to 'confirmation' in the sense of E (or, better still, C) rather than to $P(e, h) - P(e)$.

11. In view of the above discussion I venture to suggest that to look for a measure of 'the degree to which a statistical hypothesis h is confirmed by experimental evidence e' along the lines indicated in V seems to be a rather hopeless task. Any such measure would lead to the selection (for a given n) of a 'best test-statement' e. It seems, however, that the choice of the test-statement is *to some extent arbitrary* and must be left to the experimenter who will have to base his decision also on considerations outside the formal framework of statistical (or logical) theory. He will have to appraise the possible test-statements (or, equivalently, different critical regions) taking into account his risks of committing different kinds of error, as, for instance, $1 - P(e, h)$, that is the probability of his rejecting h when h is in fact true. He will perhaps consider hypotheses alternative to h. And he will assess the importance for him of avoiding one error rather another. His ultimate decision concerning e will have to depend upon factors like these. I do not consider it necessary to go into the details of such a procedure as the problem of choice of e has been elaborately described by Neyman quite a long time ago. I am well aware that I am not saying anything new here.

To put it another way: the distinction between evidence refuting (leading to rejection of) and not refuting a given statistical hypothesis seems to be of an arbitrary character and depends (also) upon considerations outside the scope of logic. No such fixed distinction seems to be indicated by the purely logical (or mathematical) relations between a given h and different possible test-statements e.

NOTES

[1] I. "Degree of Confirmation", Vol. V, No 18, 1954; II. "'Content' and 'Degree of Confirmation': A Reply to Dr. Bar-Hillel", Vol. VI, No 22, 1955; III. Carnap, Bar-Hillel, Popper: "Content and Degree of Confirmation", Vol. VII, No 27, 1956; IV. "A Second Note on Degree of Confirmation", Vol. VII, No 28, 1957; V. "A Third Note on Degree of Corroboration or Confirmation", Vol. VIII, No 32, 1958. See also: VI. "Probability Magic or Knowledge out of Ignorance", *Dialectics*, Vol. 11, No 3/4, 1957.

[2] Popper seems not unaware of the fact when he writes: "If, as in our case, h is a statistical hypothesis, and e the report of the result of statistical tests of h, then $C(h, e)$ is a measure of the degree to which these tests corroborate h, exactly as in the case of a non-statistical hypothesis. It should be mentioned, however, *as opposed to the case of a non-statistical hypothesis, it might be at times quite easy to estimate the numerical values of* $E(h, e)$ *and even of* $C(h, e)$ *if* h *is a statistical hypothesis*". (V, p. 298, italics mine.)

[3] Popper writes: "It does not matter, in the context of the present paper, whether the objective interpretation [which he adopts – K.S.] is a purely statistical interpretation" (V, p. 297). Which shows that the probabilities in V are statistical.

[4] There is a minor mistake in V, p. 299. The postulated equidistribution ought to ascribe the probability $1/n + 1$ and not $1/n$ to each possible outcome. In view of this the value of E (calculated on the following page) for a positively verified universal law ought to be $n/(n+2) = 1 - \{2/(n+2)\}$ [and not:$(n -)1/(n +)1 = 1 - \{2/(n +)\}1$]. Of course, for large n the difference is negligible.

[5] It would perhaps be better, in order to avoid confusion, to use a different notation. Thus, if we designate the conditional probability by $P(e, h)$ or $P(e/h)$ we could designate the probability of e calculated under the assumption that h is true by, say, $P_h(e)$.

[6] The expression itself seems to be suggested by a simplified form of Bayes theorem:

$$P(h, e) = P(h) \cdot \frac{P(e, h)}{P(e)}.$$

It is easy to see that if the measure of confirmation is to be an increasing function of $P(h, e)$, the absolute probability of h being kept constant, then this measure ought to be an increasing function of $P(e, h)$ and a decreasing one of $P(e)$. But I have already tried to call attention to the fact that the application of Bayes theorem implies that p is a random variable – an assumption we do not want to make.

[7] Cf. the last two paragraphs on the page 299 in V. They constitute, so far as I know, the only suggestion concerning the choice of δ if n is already fixed.

9

ON SOME BASIC PATTERNS OF STATISTICAL INFERENCE

1. Statistical inference consists in drawing conclusions about the distribution of a random variable X in a population on the basis values of X found in a random sample taken from the population in question. A rule of statistical inference is, therefore, a function associating conclusions concerning the distribution of X with particular sample-values of X. It is the task of statistical theory to formulate such rules of inference, to investigate their properties, and also to answer the question, which of them are to be considered 'good' (that is, accepted from some point of view), or possibly 'the best of all rules' with respect to a given problem (the problem itself being determined by the set of possible premises, the set of possible conclusions and the probability relation between conclusion and premises).[1]

Now, it is a well-known fact that since the appearance of Wald's [11] at least two theories of statistics were put forward. Following Luce and Raiffa [5, p. 318] I shall call them the classical and the modern theory. They single out different rules of inference as 'good' ('to be adopted') and – what is perhaps even more important from the logical point of view – they justify them in different ways. The controversy between adherents of these two theories seems to continue unabated.

It is not the purpose of this paper to take side on this issue, or even to review the argument for or against the modern theory. I shall try instead to contribute to the clarification of the issue by contrasting briefly the rules of inference recommended by the two theories, together with their respective justification. In doing this I shall restrict my discussion to the case of parametric inference with one parameter unknown and experimentation fixed. The case in question may be characterized as follows.

Suppose that the distribution of a random variable X is defined by the density function $p(x, \theta)$, θ being the parameter whose actual value is not known; the density function is well-determined except for the value of θ. The set Θ of all possible values of θ is not restricted in any way except for the (rather weak) assumption that it is bounded. The premiss of the intended inference concerning the value of θ is provided by the result of an experiment consisting in n independent random observations of the value X, say: $x_1, \ldots$.

x_n (let us designate such set of experimental data by e; the set of all possible e's is called the sample space and will be designated by E). The premiss, therefore, is of the form:

$$(X_1 - x_1) \wedge (X_2 = x_2) \wedge \cdots \wedge (X_n = x_n) \tag{1}$$

where X_i is the variable representing the i-th observation, distributed according to $p(x_i, \theta)$.

It is also assumed that θ itself is not a random variable. The justification for this assumption is that if it were a random variable with a known distribution the Bayes theorem would provide sufficient grounds for drawing conclusions concerning the value of θ; and the case of the *a priori* distribution unknown could perhaps be treated along the lines indicated by Robbins [8]. Besides, it seems that the case when θ is not a random variable is more important because much more often met in practice.

The question now is: what can be inferred from the premiss (1) concerning the actual value of θ?

2. The answer provided by the modern theory (called also 'theory of decision functions' – Wald [11], or 'theory of statistical decisions' – Blackwell and Girshick [1]) is as follows.

Each possible conclusion can be of the form $\theta = \Theta_0$ (so-called simple hypothesis) or $\theta \in \Theta_0$ where Θ_0 is a subset of Θ (so-called composite hypothesis). The set of all possible conclusions will be designated by H^*. But as only conclusions of a particular type are needed a set H of admissible conclusions is defined, where $H \subset H^*$. A rule of inference can be then interpreted as a function d defined on the set E of all possible premisses and having its values in H.[2] The problem, how to appraise such rules of inference, is then solved in the following way.

If the premiss is e and the rule adopted d then the conclusion will be $d(e)$, where $d(e) \in H$. At the same time true value of the parameter in question may be any number belonging to Θ, say θ. Of course, it would be best if θ were such that $d(e)$ be true. In general, some situations of the kind $(\theta, d(e))$ are preferred to others. Suppose that the strength of such preferences can be expressed by a function $L(\theta, d(e))$ interpreted as 'loss' (negative utility) associated with $(\theta, d(e))$. Then for every θ a rule of inference d can be characterized by the expected value of loss function L:

$$E_e[L(\theta, d(e))] = r(\theta, d) \tag{2}$$

Expected loss r (usually called 'risk') is by definition of the term 'loss', something to be minimized. And a rule of inference d will be considered good if it fulfils that purpose. As r, however, depends also on θ, the value of which is unknown, an explication is needed of the expression 'to minimize' with respect to such a situation. Several explications have been suggested and it is not my purpose to discuss them here.[3] It will be enough to say that within the conceptual framework of modern theory the adoption of a rule of inference is justified in terms of 'minimizing the risk'. Such a justification seems to be unassailable if the two assumptions characteristic of the modern theory are accepted: (1) that for every case of statistical inference a loss function can be defined in a sufficiently non-arbitrary way, and (2) that the meaning of the expression 'to minimize the risk' can be explicated satisfactorily. Both points are rather disputable and especially the first one is often questioned. It has been argued then when no practical action is made to depend on the conclusion accepted it is in many (if not all) cases difficult to associate a definite strength of preference with all possible pairs: true value of θ, conclusion accepted (cf., for instance. Fisher [3, p. 102–103]). Whether this objection applies to all cases of inference (that is: irrespective of how the set H of admissible hypotheses is defined), remains an open problem (cf. for instance [9]). A few words will be said on that point towards the end of the present paper.

3. It ought to be clear from the above exposition that modern theory covers all cases of statistical inference: perhaps its most logically important feature is that it provides a common justification for all the rules of inference recommended by it. Now, with the classical theory this is not the case. It splits up the class of all rules of inference into three parts: the rules used in the course of verifying a hypothesis concerning θ and the rules used for the purpose of point – or interval – estimating θ. And the justification seems not to be the same for the three kinds of rules. This is perhaps what makes it so difficult to compare the logical basis of the two theories. The question might, therefore, be asked: can the rules of inference recommended by the classical theory be reduced to a common form and be justified in the same way? In what is going to follow I shall try to show why this question must be answered in the negative. The basic rule of inference in the classical theory have, therefore, to be considered in the following paragraphs.

4. Let us for the moment restrict our attention to two classes of rules of inference: (A) those leading to the negation of simple hypothesis only, that is: to sentence of the type $\sim(\theta = \theta_0)$ and (B) those for which the conclusion can also be a non-refuted simple hypothesis, that is: a sentence of the type $\theta = \theta_0$. In other words: rules A allow only to reject simple hypotheses, while rules B allow also to accept them. Of course, rules B yield, in general, much stronger conclusions than rules A, since the sentence $\theta = \theta_0$ implies $\sim(\theta = \theta_1)$ for all $\theta_1 \neq \theta_0$. Rules of the type A and B are, in the classical theory, appraised in the light of different criteria and it is difficult to see how the criteria in question could be reduced to a single one. Let us consider the question in some detail.

5. There is small doubt that in the so-called tests of significance rules of the type A are exclusively used. The purpose of the tests is, as Fisher puts it, "to assess the strength of evidence that a particular hypothesis (. . .) appears to be untenable" [3, p. 92]. Thus in a test of significance a hypothesis is rejected if there is evidence which is considered to be 'almost incompatible' with the hypothesis in question. If, however, such an evidence is not there, this is not considered by the classical theory to constitute sufficient grounds for accepting the hypothesis (cf., for instance, Fisz [4, p. 228]). The reason for this will be discussed later on.

In the light of the above remarks a test of significance (with respect to the case defined in §1) can be described as follows. For each hypothesis $\theta = \theta_0$, the set of all such outcomes of the experiment is defined which are considered 'almost incompatible' with the hypothesis in question. This set, called critical region for $\theta = \theta_0$, will be designated by $K_{\theta 0}$. And the rule says that if the actual outcome e belongs to $K_{\theta 0}$ then $\theta = \theta_0$ is to be rejected. The rule of inferences thus of the form:

$$e \in K_{\theta 0} \longrightarrow -(\theta = \theta_0) \tag{3}$$

where '$\longrightarrow$' means 'therefore'. It is to be observed that the very term 'rule of inference' is used here in a wider sense than that given to it in the modern theory. It is the property of a rule of inference in the modern theory that it associates some conclusion with every premiss (the expected value of loss could be calculated otherwise). In the case under discussion no conclusion is associated with $e \in K_{\theta 0}$.

The crucial expression 'almost incompatible' is usually taken to mean that an event $e \in K_{\theta 0}$ ought to happen very rarely under the assumption that $\theta = \theta_0$. The critical region ought to fulfil the condition

$$P(e \in K_{\theta 0} \mid \theta = \theta_0) = \alpha, \tag{4}$$

where α is a small number arbitrarily chosen. The rule (3) is then closely similar to modus tollens, with the implication replaced by $1 - \alpha$ probability relation. (Czerwiński [2] calls it 'weakened modus tollens').

It ought, of course, to be noted that condition (3) is far from determining the critical region for $\theta = \theta_0$ in a unique way, and even far from excluding critical regions having very undesirable (not to say: absurd) properties. Some other conditions will have therefore to be added.

It should be noticed, however, that very often no such conditions are to be found in statistical textbooks addressing the issue of the test of significance. The method of construction of an 'appropriate' critical region is left to the 'common sense' of the experimenter. It seems that this attitude can be explained by the reluctance of some adherents of the classical theory to consider hypotheses alternative to the one being tested; it is, however, difficult to see how any general conditions to be satisfied by the critical region (such as, for instance, impartiality) could be expressed if not in terms of hypotheses alternative to the one tested. Besides, if it is assumed that the hypothesis tested can be false, it would seem hardly reasonable to close our eyes to the fact that some hypotheses alternative to $\theta = \theta_0$ do exist. In our case we assumed that the set Θ of all possible values of θ is well defined (and not restricted in any special way). I do not hesitate, therefore, to add to the account of the tests of significance a conditions imposed on the critical region, to be found usually under the heading 'verification of hypotheses'.

So it seems reasonable to postulate that $K_{\theta 0}$ be unbiased and/or that it be most powerful (if this is possible) or that it be of the Neyman type A (i.e. unbiased and most powerful for values of θ very near θ_0) etc. There is no need to go into the details of such postulates as their formulation can easily be found in many sources. What is essential from the standpoint adopted here is that in classical theory a critical region satisfying the condition (4) is associated with each hypothesis of the type $\theta = \theta_0$. Any method of constructing such critical regions provides an explication of the expression: 'the hypothesis $\theta = \theta_0$ is undermined by the outcome e of the experiment'.[4]

The usual justification of a rule of type (3) is that the probability of it's yielding a false conclusion is α, and thus arbitrarily small. This seems to be a very good justification if we wish that the probability of accepting an erroneous conclusion, on the basis of an uncertain inference be small.

On the other hand, the rule (3) may be objected to on the ground that the conclusion it provides is excessively weak (in fact, it is the weakest non-trivial conclusion possible). A way of obviating this objection would be to add a second rule, recommending acceptance of $\theta = \theta_0$ if $e \in K_{\theta 0}$:

$$e \in K_{\theta 0} \longrightarrow \theta = \theta_0 . \tag{5}$$

The sample space E is thus divided into $K_{\theta 0}$ (the region of rejection of $\theta = \theta_0$) and $E - K_{\theta 0}$ (the region of acceptance of $\theta = \theta_0$). This is the idea underlying the theory of verification of hypotheses[5] as developed by Neyman and Pearson.

It seems, however, that if no restrictions are imposed on the set Θ of all possible values of the θ, the rule (5) will be difficult to justify – even if $K_{\theta 0}$ satisfies very strong conditions. The point is that if e belongs to the acceptance region of $\theta = \theta_0$, it may as well belong to the acceptance region, say $E - K_{\theta 1}$, of some other hypothesis $\theta = \theta_1 \neq \theta_0$, defined in exactly the same way. If this is the case, then any argument in favour of accepting $\theta = \theta_0$ is by the same token an argument in favour of $\theta = \theta_0$. In other words the same e makes true the premiss of the following rule:

$$e \in K_{\theta 1} \longrightarrow \theta = \theta_1 , \tag{6}$$

where $K_{\theta 1}$ is the critical region for $\theta = \theta_1$ in exactly the same sense as $K_{\theta 0}$ is for $\theta = \theta_0$. There is no criterion inherent in the argument allowing to choose between such two (or, as the case may be, infinitely many) rules of inference yielding contradictory conclusion from the same outcome e of the experiment.

Of course, in some special cases (when, for instance, Θ is limited to two possible values of θ) the adoption of a rule of type (5) might be very well justified.

6. Rules of type (3) can easily be generalized. If for each simple hypothesis $\theta = \theta_0$ the critical region $K_{\theta 0}$ is defined, then every particular e_0 divides the set Θ into disjoint subsets: class $A(e_0)$ of all such $\theta = \theta_0$ is undermined by e_0, and the class $A(e_0)$ of all such θ_1 that $\theta = \theta_1$ is not undermined by e_0, the meaning of the expression 'undermined' being explicated by the method of constructing the critical region in question. Thus the class $A(e_0)$ consists of all such θ_0 that $e_0 \in K_{\theta 0}$. For such θ_0 the rule (3) leads to the conclusion: $-(\theta = \theta_0)$. The set Θ of all possible values of θ is thus reduced by repeated application of (3) to

$A(e_0)$. This fact can be summarized by the following rule of inference:

$$e = e_0 \longrightarrow \theta \in A(e_0). \tag{7}$$

It is easy to see that $A(e_0)$ is a confidence interval for θ associated with the outcome e_0 of the experiment. A confidence intervals has in fact been defined (for the case of one parameter unknown) by Neyman (see for instance [6], p. 197) as such subset of Θ dependent on e, say $C(e)$, that

$$P(\theta \in C(e)) = 1 - \alpha, \tag{8}$$

for each possible θ, $1 - \alpha$ being 'high' probability. It is easy to see that condition (8) is satisfied by $A(e_0)$. In fact, for each θ the probability that e will not fall into its region is $1 - \alpha$; which is at the same time the probability that $\theta \in A(e)$.[6]

Rules of inference leading to the conclusion that θ belongs to a confidence interval appear thus to be a generalization of the rules used in tests of significance to reject simple hypotheses. Their justification is, therefore, closely analogous to that of the rule (3): the probability of the rule (7) yielding a false conclusion is equal to an arbitrarily[7] small number α, no matter what the true value of θ may be (cf. Neyman [6, p. 197]).

Here again it may be said that if a rule of inference is to be appraised by the probability that using this rule one shall accept a false conclusion, then the above validation of (7) seems very good indeed. But this convincing validation of (7) is obtained only at the cost of one's being satisfied with a weak conclusion concerning the value of θ. It is true that the set $A(e_0)$ might, in some cases, be not very large. It is even true that it can be made smaller by increasing the number n of observations if the class of critical regions determining $A(e_0)$ possesses certain asymptotic properties with respect to n. The fact still remains that the conclusion of (7) is not a simple hypothesis while some definite (although hypothetical) value of θ may be needed, e.g. for some theoretical purposes.

The above considerations lead to the adoption of rules belonging to the part of classical theory known as point-estimation of θ.

7. In point-estimation, rules of inference used have always a sentence of the form $\theta = \theta_0$ as conclusion. Each such rule can, therefore, be thought of as a function called estimator, say $\theta'(e)$, defined on E and having its values in Θ. The rule can be expressed, accordingly, as

$$\theta'(e) = \theta_0 \longrightarrow \theta = \theta_0. \tag{9}$$

It is easily seen that any appraisal of rule (9) must be made in terms of the properties of the function θ' defining the rule in question. But which properties? Let us see whether the validation of (9) can be achieved in a way analogous to that of (3) or (7). The former rules were justified in terms of the probability of accepting a true conclusion (or, alternatively, the probability of accepting a false one). What is this probability for rules of type (9)?

This depends, of course, on the nature of θ' and Θ. Concerning Θ we can safely assume that it is an infinite set: otherwise the problem of estimating θ' could hardly arise. There remains θ' which is either a continuous or discrete random variable.

Let us first assume that θ' is continuous (a necessary condition for that is, of course, that X be continuous). Then the probability of accepting a true conclusion by rule (9), i.e.

$$P(\theta'(e) = \theta_0 \mid \theta = \theta_0) \tag{10}$$

is 0 for every θ_0. Thus all rules of the type (9) are equivalent in the light of the criterion used to appraise rules in tests of significance and confidence interval estimation. What is more, they are of the worst kind possible, in the sense of this criterion. If we call 'accepting a true conclusion' a 'success', then the criterion classifies a rule as 'good' when the probability of achieving a success while using the rule is high (the word 'high' having been defined arbitrarily, within certain limits). But in the case of (9) the probability of success is the lowest possible. In fact, achieving success when using (9) is hopeless, irrespective of the choice of estimator θ'.

The situation is hardly better when θ' is a discrete variable. The probability (10) is then not equal to 0 for some θ_0 (i.e. for those θ_0 which happen to be the values of θ). But for infinitely many remaining hypotheses of the type $\theta = \theta_0$ the probability of their being accepted when they are in fact true is 0.

Of course, the rule (9) can be modified so as to make the probability of premiss, given conclusion, not equal to 0. A sufficient (and necessary) condition for that would be to replace $\theta'(e) = \theta_0$ by $\theta(e) \in A$ where A is some set of non-zero measure. But then we get a rule analogous to (5) and the objection concerning (5) applies. Besides, infinitely many possible hypotheses would in that case have to be 'left out': the probability of their being accepted when true would again be 0.

Thus in point-estimation another criterion for appraising rules of inference is needed. The classical theory provides it indirectly via certain requirements concerning the estimator θ'. The most intuitive of them are: that the estimator be unbiased (11), consistent (12) and the most effective (13), if possible.

$$\text{If } \theta = \theta_0, \quad \text{then } E(\theta') = \theta_0 \tag{11}$$

$$\text{If } \theta = \theta_0, \quad \text{then for every } \epsilon > 0; \lim_{n \to \infty} P\,(|\theta' - \theta_0| > 0) = 0 \tag{12}$$

$$\text{If } \theta = \theta_0, \quad \text{then } E[(\theta' - \theta_0)^2] = \text{min.} \tag{13}$$

Usual justification for adopting such postulates is that if θ satisfies them, then: (i) by (11) and (13) the values of θ' are concentrated with high probability near the true value of the parameter, and (ii) by (12) an increase of the sample size increases the degree of this concentration.

Speaking in terms of rule (9): suppose that $\theta = \theta_0$, but $\theta' = \theta_1 \neq \theta_0$. Then we accept the conclusion $\theta = \theta_1$ which is false which is only to be expected.

But if θ' satisfies the condition (11) and (13), the probability is as high as possible that the difference between true value and the value asserted in the conclusion is within certain narrow limits: $|\theta_0 - \theta_1| \leq \varepsilon$. (The argument is not essentially changed when 'minimum variance of θ'' in (13) is replaced by 'asymptotic minimum variance'.)

Thus it appears that a significant change has been made in the criterion for appraising rules of inference. In the former case 'success' means 'acceptance of a true conclusion'. Now 'a small error in the accepted conclusion' is also considered success. In both cases a rule is considered good if there is a high probability of achieving 'success' when using it.

Such justification of the rule (9) seems also convincing if we agree that when errors are unavoidable we prefer smaller errors to greater ones. It all, of course, hinges on the fact that when the conclusion is a simple hypothesis, the size of error committed can be measured by the distance between the value of θ accepted, and its true value. The simplest such measure is $|\theta' - \theta|$ but any monotone increasing function of this difference can also fulfil that purpose (cf. Neyman [6, p. 227]).

8. To sum up. The aim of this paper was to compare the justification of rules of inference in the two theories of statistics. For this purpose a simple (though, I think, basic) model has been considered.

The result of the discussion is that in the modern theory the justification is in a sense, trivial. In this theory certain value-judgments (in the form of loss – or disutility function) are introduced as additional data. Then a rule recommended is justified in terms of those value judgments. This, incidentally, seems to be an instance of a more general thesis: that any rule of inference (as indeed any rule prescribing a certain conduct) can only be justified in terms of some purpose to be achieved by following this rule; that it will be considered good if it either guarantees the attainment of this aim or, at least, makes it highly probable. Now, if the aim, when following a rule of inference, is – by virtue of the meaning of the term 'loss' – to minimize expected loss, a rule is indeed justified when the expected loss associated with it is minimal in some sense.

This justification is quite straightforward and would perhaps seem entirely convincing, were it not for an assumption it makes use of, i.e. the loss function itself. It has often been pointed out that it may be all right for the loss function to take almost any shape when some practical considerations are involved. But inference can, in principle at least, serve no immediate practical purpose: its only purpose may simply be to obtain true statements of a certain kind. In this case the arbitrariness in the choice of loss function makes the justification of rules recommended by modern theory rather unsatisfactory. Perhaps, it is argued, arbitrariness can not be entirely eliminated when inductive rules of inference are contrasted: but it must at least be drastically cut down.

Now, in the classical theory the purpose of inference is defined in a different way. It is not assumed that preferences (always a subjective thing) are defined over the set of all outcomes of inference, meaning by 'outcome' the conjunction of conclusion and the description of true state of things. The purpose is instead defined from the outset as that of accepting a true conclusion or, at least, as 'near' to the truth as possible. This is, of course, a rather vague statement and it needs a more precise formulation.

In the above analysis an attempt was made to show, with respect to a given model, that in classical theory two classes of rules can be distinguished: those leading to a weak conclusion (i.e. of the type $-(\theta = \theta_0)$ or $\theta \in \Theta_0$) and those leading to a strong one (i.e. of the type $\theta = \theta_0$). The purpose to be achieved by following the rules of the two kinds seems not be the same. For a rule of the first kind it is: to accept a true conclusion. For the rule of the second kind it is: to accept a conclusion which, although false (this is unavoidable), does not deviate from the truth by more than a certain (small) margin. In both cases success can not be guaranteed.

But very often a rule can be constructed, such that the probability of success when following it is 'high', i.e. above certain preassigned level.[8] Such a rule can be regarded as justified with respect to the purpose it is to fulfil.

Finally, it can be remarked that in the case of rules of the second kind an obvious link exists between the classical and the modern theory. In this (and only this) case the size of error can be measured and it is the aim of the classical theory to reduce (in certain sense) this error. Thus a loss function may be said to exist, which is non-arbitrary to a high degree. The fact has been noticed long ago by Laplace, and made use of by Gauss who adopted $(\theta' - \theta)^2$ as the loss function (see Neyman [6, p. 228]).

9. I am only too well aware of the incompleteness of the above exposition. The simplifications and omissions are, in fact, so many that it would be quite a job even to enumerate them. For instance, I left out non-parametric inference. I also omitted verification of composite hypotheses. I considered only the case of one parameter unknown (from the point of view of practice this is a particularly serious limitation). Properties of critical regions (some of them have been mentioned in this paper) deserve a careful logical analysis. I was also consistently dismissing mathematical difficulties in finding rules of inference which possess the desired properties.

The purpose of the present paper was, however, to compare the justification of rules of inference in the modern theory with that in the classical one. As the classical theory consists of various and elaborate techniques difficult to reduce to a single logical basis, some rather drastic choices had to be made. I hope that the choice was a happy one.

Allatum est die 15 Decembris 1959

NOTES

[1] It is not implied that the theory of statistics reduces to answering questions of this kind. There remains, for instance, the important and often difficult problem, how to design the experiment furnishing the premisses of the inference.

[2] Actually, the theory of decision functions is more general than that and covers also the cases when any decision (possibly of a practical) nature can be made dependent on (1). In this paper, however, I am considering only the case of inference, identifying therefore a conclusion h with the decision to accept h.

[3] An excellent exposition can be found in [5]. See also [10].
[4] I use the expression 'undermined' following a suggestion by Popper in [7], p. 396.
[5] Let us notice again that we are now restricting our attention to simple hypotheses.
[6] In this expression $A(e)$ is a random variable being a function of e; θ, according to our assumption is not.
[7] Of course, if the number n of observations is kept constant, α cannot in general be decreased without, at the same time, making $A(e_0)$ larger, and as a result of this conclusion $\theta \in A(e_0)$ becomes logically weaker.
[8] A necessary condition for the possibility of constructing such a rule is usually that the number n of observations be not less than, say, n_0, where n_0 depends on the parameters of the distribution of X and on the explication of words like 'high', 'small' etc. This is the reason why asymptotic properties of critical regions and estimators play such an important role in statistical theory.

REFERENCES

[1] Blackwell D., Girshick M.A., "Theory of Games and Statistical Decisions", J. Wiley, New York 1954.
[2] Czerwiński, Z., "On the Relation of Statistical Inference to Traditional Induction and Deduction", *Studia Logica* VII.
[3] Fisher R.A., "Statistical Methods and Scientific Inference", Oliver & Boyd, Edinburgh-London 1956.
[4] Fisz M., "Rachunek prawdopodobieństwa i statystyka matematyczna", PWN, Warszawa 1954.
[5] Luce D., Raifa H., "Games and Decisions", J. Wiley, New York 1957.
[6] Neyman J., "Lectures and Conferences on Mathematical Statistics and Probability", Washington 1952.
[7] Popper K.R., "The Logic of Scientific Discovery", Hutchinson & Co., London 1959.
[8] Robbins H.: "On Empirical Bayes Approach to Statistics", *Proceedings of the Third Berkeley Symposium on Mathematical Statistics and Probability*, 1956.
[9] Szaniawski K., "Wnioskowanie czy Behaviour"?, *Studia Filozoficzne* 6 (9) 1958. [see the English translation of this paper, "Inference or Behaviour", this volume, pp. 54–61.]
[10] Szaniawski K., "Some Remarks Concerning the Criterion of Rational Decision-Making", *Studia Logica* IX. [see this volume, pp. 114–127.]
[11] Wald A., "Statistical Decisions Functions", J. Wiley, New York 1950.

10
A METHOD OF DECIDING BETWEEN N STATISTICAL HYPOTHESES

I am going to discuss briefly generalization of sequential probability ratio test; this well-known test has been designed, and its properties investigated, by Abraham Wald.[2] The (primary) purpose of Wald's method was: to test a simple statistical hypothesis against one simple alternative. Some ingenious devices have been used by Wald to extend the range of applications of the method: they all consisted in reducing the problem to the above-mentioned form, i.e. that of the choice between two simple hypotheses. It seems, however, that the method itself could be slightly generalized to cover the case of choosing between N simple hypotheses. The present paper is a preliminary discussion of such a possibility.

It is only fair to add that the proposed generalization may turn out to have no appreciable practical significance. Even if it were so, it might turn out to have some interesting properties from a purely theoretical point of view, that of the logic of uncertain inference.

We are going to assume that the functional form of the distribution of a random variable X is known to be $f(x; \theta)$,[3] where θ is a parameter the exact value of which is to be determined on the basis of experimental evidence. Let us assume further that one and only one of the hypotheses $h_1, \ldots, h_N$ concerning the value of θ is true; where h_i states that $\theta = \theta_i$. The problem now is how to decide between those hypotheses (i.e. which hypothesis to accept) on the basis of an outcome of n independent trials, the number n of trials being not fixed in advance. The desired rule of inference is thus to be a sequential nature.

Following Wald, by a sequential rule of inference (with respect to the case described above) we shall mean a rule defined in this way. For each $n(n = 1, 2, \ldots)$ the set S_n of all possible samples $\{x_1, \ldots x_n\}$ is divided into $N + 1$ disjoint subsets $R_n^1, \ldots R_n^N, R_n^{N+1}$; if the sample obtained belongs to the subset $R_n^s(s = 1, \ldots N)$ the hypothesis h_s is to be accepted; if the sample belongs to the subset R_n^{N+1} no hypothesis is accepted and $(n + 1)$th trial is made.[4]

The rule of inference I am going to formulate belongs to the class of rules just defined. It can be stated as follows:

For an outcome $\{x_1, \ldots x_n\}$ of n independent trials $(n = 1, 2, \ldots)$ accept the hypothesis h_r $(1 \leq r \leq N)$ if and only if

$$f_r(x_1) \cdot f_r(x_2) \cdot \ldots \cdot f_r(x_n) \geq A_{rs} \cdot f_s(x_1) \cdot \ldots \cdot f(x_n) \quad \text{for all } s \neq r, \tag{1}$$

where the numbers $|A_{rs}|$ are constants < 1 and the expression $f_\alpha(x)$ stands for $f(x; \theta\alpha)$. If there is no such r for which the set of inequalities (1) is satisfied by the outcome $\{x_1, \ldots x_n\}$, then no hypothesis is accepted and the $(n + 1)$th trial should be made.

It is easy to see that the above rule (let us call it R) is a generalization of the sequential probability ratio test: we obtain this test in the case of $N = 2$ by putting A for A_{21} and B for $1/A_{12}$. The rule then reduces to the form:

if $Z_n \leq B$ accept h_1
if $A \leq Z_n$ accept h_2
if $B < Z_n < A$ take $(n + 1)$th observation,

where $Z_n = f_2(x_1) \cdot \ldots \cdot f_2(x_n)/f_1(x_1) \cdot \ldots \cdot f_1(x_n)$.

The rule R may also be looked upon as a generalization of the much discussed maximum likelihood principle (m.l.) (for the case when the number of hypotheses is finite). The maximum likelihood principle recommends the acceptance of the hypothesis for which the probability (or probability density) of the outcome actually obtained is maximum. Obviously, if we put in the rule R all the constants A_{rs} equal to 1 we get the principle in question. Of course, in such a special case there is always (that is, for all n) at least one hypothesis h_r satisfying the condition (1). That is why the m.l. principle is a 'degenerate' sequential rule, i.e. such for which the number n of trials is fixed in advance; also it does not guarantee a unique solution, in view of the fact that more than one hypothesis can satisfy (1) if all the constants A_{rs} in it vanish.

Some properties of the rule R can be roughly determined in a way analogous to that followed by Wald. Let us denote by ω_r the probability that R will lead to acceptance of the hypothesis h_r when h_r is in fact true; by β_{rs} $(r \neq s)$ we are going to denote the probability that R will lead to the acceptance of h_r when h_s is true. And we shall say that $\{x_1, \ldots x_n\}$ is a sample of the type r if the relation

$$f_t(x_1) \cdot \ldots \cdot f_t(x_m) \geq A_{ts} \cdot f_s(x_1) \cdot \ldots \cdot f_s(x_m) \quad \text{for all } s \neq t, \tag{2}$$

is not satisfied for $m = 1, \ldots n - 1$, $1 \leq t \leq N$ and is satisfied for all $m = n$, $t = r$. According to the rule R, the hypothesis h_r is accepted if and only if the sample actually obtained is of the type r. But each sample of the type r satisfies the condition (1). Thus for each particular $s \neq r$, the probability (or probability density) of obtaining the sample $\{x_1, \ldots x_n\}$ of the type r is at least A_{rs} times as great under the assumption that h_r than under the assumption that h_s. Therefore, the probability measure of all samples of the type r under the assumption that h_r must be at least A_{rs} times their probability measure under the assumption that h_s. We thus have:

$$\omega_r \geq A_{rs} \cdot \beta_{rs} \quad \text{for all } s \neq r. \tag{3}$$

Hence:

$$\beta_{rs} \leq 1/A_{rs} \quad \text{for all } s \neq r. \tag{4}$$

The inequality (4) provides an upper bound for the probability of an error consisting in accepting the hypothesis h_r when some other hypothesis h_s is in fact true. Obviously, this upper bound is set to high: it is enough to notice that in deriving (4) from (3) we have taken advantage of the fact that ω_r cannot be greater than 1, while actually ω_r is always lower than 1. This is important from the point of view of practical applications: if we want to keep the probabilities of errors within certain limits and to have the expected number of trials not greater than it is necessary to achieve this purpose (because of the cost of experimentation), then a more exact relation between β_{rs} and A_{rs} is needed. However, for the limited purpose of this paper it is enough to remark that p_{rs} depends upon A_{rs} and by the proper choice of this constant it can be made as small as desired.

We are coming now to the question of ω_{rs}, that is: the probability of accepting h_r when h_r is true. Let us consider first the probability, say α_r, of rejecting h_r when h_r is true. In view of (4) we have:

$$\alpha_r = \textstyle\sum_{s \neq r} \beta_{sr} \leq \sum_{s \neq r} 1/A_{sr}. \tag{5}$$

Suppose now that the probability (under the assumption that h_r is true) that the procedure will terminate in a finite number of trials is 1. We would then have:

$$\omega_r = 1 - \alpha_r \geq 1 - \textstyle\sum_{s \neq r} 1/A_{sr}, \tag{6}$$

and by a suitable choice of the constants A_{sr}, ω_r could be made as high as desired. But this last assumption has not been proved so far. Before we proceed to the discussion of this point it should be remarked that – irrespective of the problem of controlling ω_r – something of that sort ought to be proved if the rule R is to have some (even if only theoretical) value. It ought to be shown that, whichever of the hypotheses $h_1, \ldots h_N$ is true, the probability is 1 that the rule R will lead to the acceptance of some hypothesis in a finite number of trials. Otherwise the possibility would be open that the rule R will fail as a means of arriving at a decision.

Let us remark that in order to prove:

> The probability is 1 that there is such m that the procedure terminates after m trails, it is sufficient to prove: (7)

> The probability that the procedure terminates after not more than n trials tends to 1 as n increases indefinitely. (8)

In order to see this, let us write A_m for the sentence 'the procedure terminates after m trials'. (7) and (8) reduce then, respectively, to

$$P[(\exists_m)A_m] = 1, \tag{9}$$

and

$$\lim_{n\to\infty} P[(\exists_m)(m \leq n \wedge A_m)] = 1, \tag{10}$$

or, to apply an obvious abbreviation:

$$P(Z) = 1 \tag{9'}$$

$$\lim_{n\to\infty} P(S_n) = 1. \tag{10'}$$

But obviously, for each n: S_n implies Z. Therefore, for each n:

$$P(Z) \geq P(S_n). \tag{11}$$

Suppose now that (9′) is not true, i.e.

$$P(Z) = q < 1. \tag{12}$$

But from (10′) it follows that for the some n_0

$$P(S_{s0}) > q, \tag{13}$$

which in view of (11) entails:

$$P(Z) > q. \tag{14}$$

We have thus arrived at a contradiction which shows that (10′) and the negation of (9′) are incompatible. Therefore (9′) follows from (10′).

It is one of our assumptions that exactly one of the hypotheses $h_1, \ldots h_N$ is true. Without any loss of generality we shall assume that the true hypothesis is h_r. By definition of the rule R, a sufficient condition for the procedure to terminate in no more than n steps is that the relation

$$f_r(x_1) \cdot \ldots \cdot f_r(x_n) \geq A_{rs} \cdot f_s(x_1) \cdot \ldots \cdot f_s(x_n) \tag{15}$$

be satisfied by the sample $\{x_1, \ldots x_n\}$ for all $s \neq r$ if the probability (under the assumption that h_r) that the relation (15) is satisfied tends to 1 as n increases indefinitely, then the probability that all the relations of the type (15) – i.e. for all $s \neq r$ – are satisfied, also tends to 1 as n increases indefinitely. Therefore, in order to prove (8) (and thus (7)) it is enough to show that for any $s \neq r$ and under the assumption that h_r is true:

$$\lim_{n\to\infty} P[f_r(x_1) \cdot \ldots \cdot f_r(x_n) \geq A_{rs} \cdot f_s(x_1) \cdot \ldots \cdot f_s(x_n)] = 1. \tag{16}$$

I am now going to prove (16) for two, admittedly most important, cases. (A) Suppose that the hypotheses $h_1, \ldots h_N$ ascribe definite values to the parameter p in binomial distribution, i.e. hypothesis h_i states that $p = p_i$. (16) reduces now to the following form. Under the assumption that $p = p_r$:

$$\lim_{n\to\infty} P[p_r^k \cdot (1 - p_r)^{n-k} \geq A \cdot p_s^k \cdot (1 - p_s)^{n-k}] = 1, \tag{17}$$

where k is the number of successes in n independent trials, $p_s \neq p_r$ and A is a constant ≥ 1 (the subscript in A_{rs} can now be dropped).

From the law of great numbers (and in the view of the assumption that $p = p_r$) we have:

$$\lim_{n\to\infty} P(|k/n - p_r| < \varepsilon) = 1, \tag{18}$$

where ε is any positive constant. It is therefore sufficient to prove that:

$$|k/n - p_r| < \varepsilon \tag{19}$$

implies

$$p_r^k(1 - p_r)^{h-k} \geq Ap_s^k(1 - p_s)^{h-k}. \tag{20}$$

But (20) is equivalent to

$$k/n \log p_r + (1 - k/n)\log(1 - p_r) - k/n \log p_s - (1 - k/n)\log(1 - p_s) \geq 1/n \log A. \tag{21}$$

from (19) we have:

$$P_r - \varepsilon < k/n < p_r + \varepsilon. \tag{22}$$

Making appropriate substitutions in (21) we now get

$$p_r \log p_r + (1 - p_r)\log(1 - p_r) - p_r \log p_s - (1 - p_r)\log(1 - p_s) - \varepsilon \times [\log p_r + \log(1 - p_r) + \log p_s + \log(1 - p_s)] \geq (1/n)\log A. \tag{23}$$

As, however, ε is a small as we wish, and $(1/n)\log A \to 0$, it is sufficient to prove that

$$\varphi(p_r, p_s) = p_r \log p_r + (1 - p_r)\log(1 - p_r) - p_r \log p_s - (1 - p_r)\log(1 - p_s) > 0. \tag{24}$$

It is easy to check that $\varphi(p_r, p_s)$ has minimum for $p_r = p_s$, and this minimum is 0. In view of the assumption that $p_r \neq p_s$ this completes the proof.

(B) Suppose now that the hypotheses $h_1, \ldots h_N$ ascribe definite values to the mean μ in a normal distribution with known variance $\sigma,^2$ i.e. a hypothesis h_i states that $\mu = \mu_i$. (16) then reduces to the following form. Under the assumption that $\mu = \mu_r$:

$$\lim_{n\to\infty} P[n_r(x_1) \cdot \ldots \cdot n_r(x_n) \geq A \cdot n_s(x_1) \cdot \ldots \cdot n_s(x_n)] = 1, \tag{25}$$

where $n_r(x)$, $n_s(x)$ are normal densities with respective means μ_r, μ_s $(\mu_r \neq \mu_s)$ and the standard deviation σ; A is a constant <1. Taking logarithms and dividing by n we obtain:

$$\lim_{n\to\infty} P\left[\frac{1}{n}\sum_{i=1}^{n} \log n_r(x_i) - \frac{1}{n}\sum_{i=1}^{n} \log n_s(x_i) \geq \frac{1}{n} \log A\right] = 1. \tag{26}$$

But from Tchebysheff's law of great numbers we have

$$\lim_{n\to\infty} P\left(\left|\frac{1}{n}\sum_{i=1}^{n}\log n_r(x_i) - C_r\right| < \varepsilon\right) = 1 \tag{27}$$

and

$$\lim_{n\to\infty} P\left(\left|\frac{1}{n}\sum_{i=1}^{n}\log n_s(x_i) - C_s\right| < \varepsilon\right) = 1, \tag{28}$$

where (in view of the assumption that $\mu = \mu_r$):

$$C_s = E[\log n_r(x)] = \int_{-\infty}^{\infty} n_r(x)\log n_r(x)\,dx \tag{29}$$

and

$$C_s = E[\log n_s(x)] = \int_{-\infty}^{\infty} n_r(x)\log n_s(x)\,dx. \tag{30}$$

In view of the fact that ε is as small as we wish and $(1/n)\log A \to 0$, it is, therefore, sufficient to prove that $C_r > C_s$. But

$$C_r = \int_{-\infty}^{\infty} n_r(x)\left[\log\frac{1}{\sigma\sqrt{2\pi}} - \frac{(x-\mu_r)^2}{2\sigma^2}\right]dx = \log\frac{1}{\sigma\sqrt{2\pi}} - \frac{1}{2} \tag{31}$$

and

$$C_s = \log\frac{1}{\sigma\sqrt{2\pi}} - \frac{1}{2\sigma^2}\int_{-\infty}^{\infty}(x-\mu_s)^2 n_r(x)\,dx$$

$$= \log\frac{1}{\sigma\sqrt{2\pi}} - \frac{1}{2\sigma^2}(m_r^{(2)} - 2\mu_r\mu_s + \mu_s^2) \tag{32}$$

where $m_r^{(2)}$ is the second moment of $n_r(x)$. Thus

$$C_r - C_s = \frac{1}{2}\left[\frac{1}{\sigma^2}(m_r^{(2)} - 2\mu_r\mu_s + \mu_r^{(2)}) - 1\right]$$

$$= \frac{1}{2\sigma^2}[m_r^{(2)} - 2\mu_r\mu_s + \mu_s^2 - (m_r^{(2)} - \mu_r^2)] = \frac{(\mu_r - \mu_s)^2}{2\sigma^2} \tag{33}$$

Therefore $C_r - C_s > 0$, because of the assumption that $\mu_r \neq \mu_s$; which completes the proof. To sum up. A sequential rule R of inference has been formulated. The premiss of such an inference is the report on the outcome of n independent trials, the number n not being fixed in advance. The conclusion is one of N simple statistical hypotheses. The rule R is a generalization of the sequential probability ratio test, as defined by Wald. At the same time in turns out to be a generalization of the maximum likelihood principle (in the case when the number of hypotheses is finite).

Some properties of the rule R have been roughly ascertained. It turns out that (as it is the case for the sequential probability ratio test) the probabilities of errors can, in principle, be made as small as desired by a suitable choice of constants in R. For two classes of hypotheses it has been proved that the rule R will lead with probability 1 to a decision in a finite number of steps.

It ought to be stressed, however, that if the rule R is to have any practical significance, its properties ought to be investigated much closer than that. The expected number of trials is particularly important in this context. Its estimation seems to be a rather difficult task.

11
A PRAGMATIC JUSTIFICATION OF RULES OF STATISTICAL INFERENCE

1. In the so-called inductive (or mathematical) statistics rules of inference of various kinds are formulated. The fundamental class of such rules is known under the technical name of rules of parametric inference with one parameter unknown, experimentation fixed, no *a priori* distribution assumed. They can be described as follows.

The conclusion states something concerning a numerical characteristic, say ω, of a population. In either specifies the extract value of this parameter (the so-called simple hypothesis; its form is: $\omega = \omega_0$), or it specifies a class of numbers, usually an interval, to which this value belongs (a so-called composite hypothesis; its form is: $\omega = \Omega$). The statement $\omega = \omega_0$ (resp. $\omega = \Omega$) can, therefore, be identified from the formal point of view with the number ω_0 (resp. with the interval Ω), and it is possible to speak of certain numbers (resp. intervals) as if they were statements.

The premiss is a report concerning the outcome of an experiment consisting of n random draws from the population in question. The outcome x (it in fact is an n-tuple of numbers) can therefore be treated as the premiss.

It is further assumed that for each value of ω the probability distribution of outcomes is completely defined; we shall denote the distribution function by $p(x \mid \omega)$. This is, of course, a shorthand notation, p being actually joint probability distribution (or density) function of n variables.

The class of rules of inference considered is thus characterized by the following data:

$$\langle X, \underline{\Omega}, p, H \rangle,$$

where X is the class of outcomes of the experiment; $\underline{\Omega}$ is the class of values of ω, assumed to be an interval, possibly unbounded; H is the class of admissible conclusions. An element of H is either a point in $\underline{\Omega}$, or interval in $\underline{\Omega}$, or a complement of a point (or of an interval) in $\underline{\Omega}$. H is defined in such a way that the following relation holds: $U_{h \in H} h = \underline{\Omega}$. In view of this, it might be said that each possible value of ω makes some hypothesis in H true, and any hypothesis in H can be true.

A rule of inference d associates a conclusion with each possible premiss. It is, therefore, a function defined on X and having its values in H. The problem now is, why are we to follow any such rule?

A well-known answer to this question has been given by the theory of decision functions. According to this theory, the use of a rule is justified if the rule maximizes, in a certain sense, the expected utility associated with decisions made according to the outcome of the experiment. Against this approach the objection is often raised that it is not clear how utility should be defined in the case of 'pure' inference, i.e. when no practical action is contemplated at the time the inference is made. It would not be possible to make it depend upon material losses and gains from actions that *could* be based on the conclusion because it is not even possible to enumerate all such actions, let alone to appraise the losses that result from them. An answer to this objection would be provided if a utility function was defined, independently of any practical use the conclusion could be put to. Hempel proposes to call it 'epistemic utility'.[5] This can be done if we assume that inferences are (in some cases at least) made in order to achieve a theoretical end. Epistemic utility could then be identified with the degree of this end's realization, and a rule of inference would be appraised in terms of such utility. In what is going to follow we shall try to make the last few statements more precise.

2. It seems clear that to assert a statement is to behave in a certain way, provided the word 'behaviour' is taken in a sufficiently wide sense. Therefore, a rule of inference is a method, if by 'method' we mean a rule making behaviour conditional upon circumstances; in the case of rules of inference the circumstances are experimental data described in the premiss.

Furthermore, it can be said that at least one way of justifying the use of a method is to show that it is not less efficient than other methods available, with respect to the end to be realized by its use. If we adopt this approach to the problem of justification of rules of inference, two questions ought to be answered:

A. What is the (theoretical) end to be achieved by means of an inference?

B. How to measure the efficiency of a method (in particular, of a rule of inference) with respect to a given end?

We shall consider these two questions in turn.

3. A definition of an end to be achieved by a possible action is always relative to a set, say E, of events that may possibly happen when this action is taken. Of course, it is impossible to enumerate, and even to imagine, all the consequences of an action. So the set E is rather arbitrarily cut out of the class of all those consequences: we say that we are not interested in other possible happenings, that they are not relevant to the purpose we have in mind. Then the definition of the end consists in answering for each $e \in E$ the following question: to what degree will the end be achieved if e happens? This question is considered answered if a function r, bounded from above, is defined on E, and by definition: $r(e) = r_0$ means: 'when e happens, the end is achieved to the degree r_0'. Let us denote the lowest upper bound of r by u. Then: if e happens and $r(e) = u$, we shall say that the end has been achieved completely.

It should be stressed that we shall need the concepts just introduced only for the purpose of *comparing* different methods as to their efficiency with respect to a given end. Therefore, it is not the absolute values of $r(e)$ that matter but rather relative distances from the full realization of the end. The choice of a unit and of a point of origin is thus immaterial, and the function r ought to be treated as an interval scale defined on E, just like in the theory of utility.

Now, what is E in our case? It has been assumed that, very loosely speaking, the purpose of an inference is a cognitive one. To what degree it is realized depends upon which conclusion has been accepted, and what is the actual value of ω. It follows that $E = \underline{\Omega} \times H$, and r is thus a function of two variables: ω and h. It is clear that $r(\omega, h)$, the degree of the end's achievement in the case of statistical inference, is what we have called earlier 'epistemic utility'. We shall postpone for the moment the discussion concerning the exact shape of $r(\omega, h)$.

4. If an inference rule d makes h depend upon x, then the degree of the end's achievement, $r(\omega, d(x))$, becomes a random variable with the probability distribution induced by $p(x \mid \omega)$. Clearly, the efficiency of d should be defined in terms of some characteristics of this distribution. The simplest possibility is to identify the efficiency of d with the expected value of the degree of the end's achievement when d is used. Of course, some other characteristics could also be introduced, such as a measure of dispersion. This would be theoretically preferable but not very realistic, since it would make for serious computational complications. Let us denote the efficiency of d, relative to ω, by $e(d \mid \omega)$. Then[6]

$$e(d \mid \omega) = {}_{Df} E[r(\omega, d(x))] = \sum r(\omega, d(x)) \cdot p(x \mid \omega).$$

Thus for a fixed ω, rules of inference can be compared as to their efficiency. We have assumed, however, that no information concerning the actual value of ω was available, except that provided by the experiment. If there is no such rule *for all* ω, its efficiency is not smaller than that of any other rule, then two approaches are possible.

A. To adopt an index, say $e(d)$, defined for the set of numbers: $e(d \mid \omega)$ (ω ranging over $\underline{\Omega}$), and depending on them in a non-decreasing way. We thus define the efficiency of d, independent of the actual value of ω. Several possible definitions of this sort are known; they will not be discussed here.[7]

As has been mentioned before, the use of particular rule d is then justified by showing that its efficiency in a specified sense is maximum in the class, say D, of all possible rules.[8] An obvious weakness of this approach consists in a somewhat arbitrary character of the index $e(d)$.

B. Let us restrict the search for the most efficient rule to a subclass, say D^*, of the class D of all rules. (In verification of hypotheses, for instance, D^* may consist of all rules having an arbitrarily chosen level of significance). In many cases there is then a rule, say d_0, which is most efficient in D^* for all ω.

This approach is often adopted in practice for the additional reason that finding a d with the highest index of efficiency, $e(d)$, is quite often extremely difficult. Its merit is that d_0 is *unquestionably* most efficient within the class D^* (in game-theoretical parlance, it weakly dominates all other rules in D^*). On the other hand, the arbitrariness (unavoidable, it seems) consists now in the choice of D^*.

5. It remains to define the end which statistical inference is to achieve, i.e. the function $r(\omega, h)$. For several reasons, the definition cannot be the same for all kinds of statistical inference. We thus have to distinguish between basic types of inferences.

Statistical inferences, as defined at the beginning of the present paper, can be roughly divided into two types, according to the character of the class H of admissible conclusions. The class H is, in turn, determined by the kind of problem the inference is intended to provide an answer to.

The problem of choice: given a set of hypotheses (it is assumed that at least one of them, but not necessarily exactly one, is true), we have to choose a hypothesis having some desirable property (e.g. truth).

The problem of verification: given a hypothesis (usually called the null hypothesis), we have to answer the question, whether this hypothesis is true or not.

The above classification is based on the distinction, first made by K. Ajdukiewicz,[9] between two types of questions: the 'which' questions, and the 'whether' questions. It also easy to see that from the formal point of view the difference between the two types of problems is the following one. If the class H has more than two elements, the inference is intended to solve the problem of choice; if H exactly two elements, it deals with the problem of verification.

Among the inferences of the first type, two particularly important cases can be singled out. When all the hypotheses are simple, $H = \underline{\Omega}$, the inference is called point estimation. When H is the class of all intervals in $\underline{\Omega}$ (the hypotheses are then composite) the inference is called interval estimation.

In the case of verification of hypotheses we distinguish between the case when null hypothesis is simple, and the case when it is composite.

Let us discuss now the definition of ends to be achieved by inferences of the above four, admittedly most important, types.

6. Point estimation. As we have already said, the class H of admissible conclusions is in this case identical with the class of possible values of the parameter. It is tempting to assume the end to be acceptance of the true hypothesis, i.e.[10]

$$r(\omega, h) = \begin{cases} 1 & \text{if } \omega = h \\ 0 & \text{if } \omega \neq h. \end{cases}$$

This, however, would lead to undesirable consequences. It is easy to see[11] that for all d:

$$e(d \mid \omega) = 0 \begin{cases} \text{for all } \omega \text{ if the set } X \text{ is continuous} \\ \text{for almost all } \omega \text{ if the set } X \text{ is discrete.} \end{cases}$$

The efficiency of any rule is thus minimum with respect to such an end: they are all equally bad means of achieving it. We are, therefore, forced to appraise the rules of point estimation from the point of view of some other end.

It is not difficult to see what this other end could be. Obviously, if the accepted value h of the parameter differs but slightly from its true value ω, we do not consider our conclusion a total failure; and the closer we get to the actual value of ω, the better. The degree of the end's realization thus depends on the difference between h and ω (the 'size error'), and the end itself is, loosely speaking, to arrive at a conclusion with a small error. The size of error can be measured by $|\omega - h|$; for computational reasons the 'Gaussian loss function', $(\omega - h)^2$, is usually preferred.

This suggests some postulates concerning $r(\omega, h)$. For all ω: $r(\omega, h)$ should increase as $|\omega - h|$ decreases; it should reach its maximum when $\omega = H$; it should be symmetrical about ω. The last requirement is justified by the fact that when the purpose of inference is a purely theoretical one, it is only the size of error that matters, and not, e.g., its sign. A simple function satisfying the above requirements can be defined by means of the 'Gaussian loss':

$$r(\omega, h) = -(\omega - h)^2.$$

The use of the rule d_0, if it exists, such that for all ω:

$$E_x[-(\omega - d_0(x))^2] = \max_{d \in D}$$

is then justified by the fact of its being most efficient with respect to the end as defined above. In statistical literature such rules ('estimation') are considered the best. Since, however, most efficient rules do not often exist, some weaker properties are postulated in place of the last mentioned. They will not be discussed here.

7. Interval estimation. The class H of admissible conclusions consists now of all intervals in $\underline{\Omega}$.

Again, let us first suppose that the end is: to derive a true conclusion. Then, of course, the most efficient rule (for all ω), say $d^*(x) = \underline{\Omega}$ for all x. Obviously, this would be satisfactory, as the conclusion $\omega \in \underline{\Omega}$ is analytic, in view of the definition of $\underline{\Omega}$. And even if $\underline{\Omega}$ were excluded from the class of admissible conclusions, it would be easy enough to find similar rules, highly efficient and leading to an almost trivial conclusion.

It thus appears that what we really want in interval estimation is a conclusion Ω which would be at the same time true and short (in other words, a true and precise estimate of ω). But these are two *different* ends, and conflicting ones at that; in the sense that the more efficient a rule is with respect to the first end, the less efficient it is with respect to the second one, and vice versa. The two ends we are speaking about can be defined as follows:

$$r_1(\omega, \Omega) = \begin{cases} 1 & \text{if } \omega \in \Omega \\ 0 & \text{if } \omega \bar{\in} \Omega, \end{cases}$$

$$r_2(\omega, \Omega) = -|\Omega| \quad \text{for all } \omega,$$

where Ω is any interval in $\underline{\Omega}$, $|\Omega|$ denotes its length, and $r_i(\omega, \Omega)$ means the degree of the i-th end's achievement when the actual value of the parameter is ω and the conclusion states that the value belongs to Ω.

Any rule of inference in interval estimation will have to be appraised in terms of those two ends. There would be no point in mixtures of r_1 and r_2, because the value of r_2 depends upon the choice of unit in measuring the length of intervals. So the approach usually adopted is the following one.

It is postulated that the efficiency of d with respect to the first end be constant for all ω and equal to $1 - \alpha$, where α is a small fraction, arbitrary chosen:

$$e_1(d \mid \omega) = E_x[r_1(\omega, d(x))] = 1 - \alpha, \quad \text{for all } \omega.$$

Let us designate by D_α the class of rules satisfying this condition.[12] Then a rule, say d_0, is sought, which would be most efficient (for all ω) in D_α with respect to the second end:

$$e_2(d_0 \mid \omega) \geq e_2(d \mid \omega), \quad \text{for all } \omega, \text{ and for all } d \in D_\alpha.$$

The use of d_0 is considered to be justified because of the fact that it has a constant and arbitrarily chosen (therefore, high) efficiency with respect to the first end, and among such rules it is most efficient (for all ω) with respect to the second one.

Of course, the order could, in principle, be reverted: e_2 postulated to be constant and e_1 maximized. Or again, it may be postulated that for a fixed unit in $\underline{\Omega}$ the length of the interval $d(x)$ does not exceed, for all x, some arbitrarily chosen number, $e_1(d \mid \omega)$ being constant. Such variations are often interesting and practically important, but they do not bring anything essentially new to the problem of justification of respective rules of inference.

8. Verification of a simple hypothesis. The hypothesis to be verified states that $\omega = \omega_0$. The class H of conclusions consists then of two elements: ω_0 and $\underline{\Omega} - \omega_0$. To simplify the notation, let us write ω_0 for $\underline{\Omega} - \omega_0$ (the negation of ω_0).

Any rule d is, of course, equivalent to a dichotomous partition of X into, say, X_d and $\bar{X}_d$ such that

$$X_d = (x)[d(x) = \omega_0], \quad \bar{X}_d = X - X_d.$$

X_d is thus the set of all such outcomes of the experiment which by the rule d lead to the acceptance of h.

Let us suppose again that the end is to derive the true conclusion, i.e.

$$\left.\begin{array}{l} r(\omega_0, \omega) = r(\omega, \omega_0) = 1 \\ r(\omega, \omega_0) = r(\omega_0, \omega_0) = 0 \end{array}\right\} \quad \text{for all } \omega \neq \omega_0,$$

where, e.g., $r(\omega, \omega_0)$ means: the degree of the end's achievement when the actual value of the parameter is ω and the null hypothesis is accepted.

We now have the following difficulty. If for any fixed x, $p(x \mid \omega)$ is a continuous function of ω, then for any d:

$$e(d \mid \omega_0) \approx 1 - e(d \mid \omega)$$

for any ω sufficient close to ω_0. In other words, the higher the efficiency of d when the null hypothesis is true, the lower its efficiency when the null hypothesis is 'nearly true'. This would force us to adopt highly paradoxical rules of inference if they were to be appraised by means of an index of efficiency we have spoken of at the beginning of the present paper.[13] Therefore the second approach is adopted.

It is postulated that the efficiency of d, under the assumption that the null hypothesis is true, be $1 - \alpha$, where α is a small fraction, arbitrarily chosen:

$$e(d \mid \omega_0) = 1 - \alpha.$$

Then the rule d_0 is adopted, if it exists, dominating in efficiency all d in the class D_α of rules satisfying the above condition:

$$e(d_0 \mid \omega) \geq e(d \mid \omega) \quad \text{for all } \omega \text{, and for all } d \in D_\alpha .$$

In statistical literature such a rule is called, uniformly most powerful at the $1 - \alpha$ level of significance. If d_0 does not exist, some weaker conditions are postulated.

But the fact still remains that for all ω sufficiently close to ω_0, the efficiency of d_0 is very low (approaching α). It might be argued that this is a consequence of an inadequate definition of the end: that if the actual value of the parameter is in a sufficiently close neighborhood of ω_0, it does not matter if we (wrongly) accept the null hypothesis. This leads us to a redefinition of r:

$$\left.\begin{aligned} r(\omega_0, \omega_0) = r(\omega, \omega_0) = 0 \\ r(\omega_0, \omega_0) = r(\omega, \bar{\omega}_0) = 1 \end{aligned}\right\} \quad \text{for all } \omega \text{ such that } |\omega - \omega_0| > c$$

$$r(\omega, \omega_0) = r(\omega, \omega_0) = 1 \quad \text{for all } \omega \text{ such that } |\omega - \omega_0| \leq c;\ \omega \neq \omega_0 .$$

In the above definition, c is an arbitrarily chosen number determining such a neighborhood of ω_0 that if the actual value of ω lies in this neighborhood, the acceptance of ω_0 is considered to be no error. Then for any d, $e(d \mid \omega) = 1$ for any ω in this neighborhood; whence the expression 'indifference zone', used in the literature.[14] It is easy to see that if the end is so defined, the undesirable consequence we have just spoken of disappears. For if the actual value of ω lies outside the indifference zone, the efficiency of d can be appreciably higher than α, provided c is suitably chosen. We can say that the end now is to obtain the true conclusion when the actual value of ω is not in the indifference zone. It is also clear that the method can be refined if we make $r(\omega, \omega_0)$ depend in a continuous way on the distance of ω from ω_0, i.e. on the size of error committed when accepting the null hypothesis.[15]

9. Verification of a composite hypothesis. The class H of admissible conclusions consists now of Ω_0 and $\underline{\Omega}_0 = \Omega - \Omega_0$, where Ω_0 is an interval, and the

null hypothesis states that $\omega \in \Omega_0$. It can again be assumed that the end is to derive the true conclusion, i.e.

$$r(\omega, \Omega_0) = \begin{cases} 1 & \text{if } \omega \in \Omega_0 \\ 0 & \text{if } \omega \bar{\in} \Omega_0 \end{cases}$$

$$r(\omega, \overline{\Omega}_0) = \begin{cases} 1 & \text{if } \omega \bar{\in} \Omega_0 \\ 0 & \text{if } \omega \in \Omega_0 . \end{cases}$$

The justification of a rule of inference with respect to this end is closely analogous to that in the case of a simple hypothesis. There is no need to reconstruct it here. It is also possible to redefine the end to be realized, by means of the 'indifference zone' method.

To sum up, in this paper I have suggested a possible justification for accepting a conclusion drawn from some experimental evidence, according to a rule of statistical inference. The expression 'statistical inference' has been limited to inferences of the simplest type.

The justification proposed I call 'pragmatic', for it treats rules of inference as rules prescribing certain purposive behaviour. The justification for following any such rule consists in showing it to be not less efficient than other possible rules, with respect to a given end.

Efficiency has been identified with the expected value of the degree of the end's achievement. In the case of rules of statistical inference it always depends on the actual, unknown, value of a parameter. For the purpose of comparing the efficiency of different rules, irrespective of the value of the parameter, either some index of efficiency is adopted, or the class of all possible rules is reduced in such a way that there exists a rule dominating in efficiency all other rules.

As to the end to be achieved by means of inference, it has been assumed that it is, broadly speaking, of a cognitive nature. In particular, no gains or losses from a contemplated practical decision based on the conclusion of inference, have been taken into account. In order to define the end more precisely, statistical inferences have first been classified into those intended to solve a problem of choice, and those leading to the solution of a problem of verification; some further subdivisions have been made, in conformity with statistical practice. Then the question of how to define the (cognitive) end has been discussed for each type of inference rule. It is only fair to add that nothing essentially new has been said above. The present paper is rather an attempt to unify and perhaps make more clear the justification actually provided by statisticians for the use of certain rules of inference.

NOTES

[1] "This much is clear: the utilities should reflect the value or disvalue which the different outcomes have from the point of view of pure scientific research rather than the practical advantages or

disadvantages that might result from the application of an accepted hypothesis, according as the latter is true or false. Let me refer to the kind of utilities thus vaguely characterized as *purely scientific*, or *epistemic*, utilities" (Carl G. Hempel, 'Inductive Inconsistencies', *Synthese*, Vol. XII, 1960, No 4, p. 465.

[2] It should be remarked that in the definition below we use a shorthand notation again: as the outcome of the experiment is an n-dimensional random variable, the sign of addition stands for multiple addition (or integration) with respect to n variables.

[3] Cf. for instance, K. Szaniawski, 'Some Remarks Concerning the Criterion of Rational Decision Making', *Studia Logica*, IX, 1960. (see this volume, pp. 114–127.]

[4] It would not, in general, be possible to justify a rule by showing that its efficiency is above a certain preassigned level, because of the fact that this level can always be raised (or lowered) by suitably redefining r.

[5] K. Ajdukiewicz, 'Zdania pytajne' [Interrogative Sentences] in *Język i poznanie* [*Language and Knowledge*], PWN, Warszawa 1960. Reprinted from *Logiczne podstawy nauczania*, 1938. See also D. Harrah, 'A Logic of Questions and Answers', *Philosophy of Science*, Vol. 28, 1961, No 1.

[6] In such a case the end has obviously the 'all or nothing' character, and we can assign any two numbers to its degrees of realization. It is most convenient to choose 0 and 1 for this purpose. The efficiency of a rule is then a number in the interval (0, 1).

[7] Cf. K. Szaniawski, 'On Some Basic Patterns of Statistical Inference', *Studia Logica*, XI, 1961. [see this volume, pp. 70–79.]

[8] The conclusion obtained by means of such a rule is called in statistical literature a confidence interval at $1 - \alpha$ level. The method has been introduced by J. Neyman.

[9] Thus, e.g., the maximum most efficient d would be the one with $e(d \mid \omega_0) = \frac{1}{2}$, and the Laplace most efficient d would lead to the rejection of ω_0, whatever the outcome of the experiment.

[10] See A. Wald, *Sequential Analysis*, J. Wiley, New York 1947, p. 27–29.

[11] Ibidem.

12

ON SEQUENTIAL INFERENCE

1 INTRODUCTION

The idea of sequential inference is probably as old as that of induction itself. All traditional accounts of induction by simple enumeration say roughly this: elements of a class A are examined, one by one, for a property B; the process is continued until either (1) an element of A turns out to be not-B (in which case the hypothesis 'all A are B' is rejected), or (2) the inference maker is satisfied that he has accumulated enough evidence to accept the hypothesis 'all A are B'. In spite of its crude form (no criterion of sufficient evidence is provided), the above procedure has certain properties characteristic of sequential inference. The accumulation of evidence on which the conclusion is to be based proceeds in consecutive steps; their number is not determined in advance.

Like all inductive procedures, sequential inference can exactly be described in probabilistic language. It is, therefore, not surprising that both the concept and the term to denote it were first introduced within the theory of mathematical statistics. The definition, due to Wald (1947), is as follows.[16]

Let $Z = \{h_1, h_2, \ldots, h_N\}$ be complete set of pairwise exclusive hypotheses. Let $X_1, X_2, \ldots$ be a series of random variables, representing the outcomes of consecutive experiments. Let $e_n = \langle x_1, x_2, \ldots, x_n \rangle$ denote a specific series of outcomes of the first n experiments; the class of all e_n will be symbolized by E_n. Now, a sequential rule of inference consists in defining for all $n (n = 1, 2, \ldots)$ a partition of E_n into $N + 1$ disjoint subsets $S_n^1, S_n^2, \ldots S_n^{N+1}$ (some of them may be empty). They have the following interpretation. If the outcome e_n actually obtained, it belongs to $S_n^k (1 \leq k \leq N)$ and the hypothesis h_k is accepted; if $e_n \in S_n^{N+1}$, no conclusion is drawn and $(n + 1)$th experiment is performed.

As Wald himself pointed out, the above is a generalization of the 'ordinary' rule of inductive inference, i.e. one with the number n of experiments fixed in advance. Such a rule is obtained if we assume $S_n^{N+1} = \varnothing$.

Having defined the concept of sequential inference, Wald gave an example of such a procedure, viz. the sequential probability ratio test. A generalization of that rule will be discussed below.

Wald's research was motivated mainly by practical needs: sequential techniques make possible to reduce the average size of experimental data, without impairing the conclusion's degree of validity. It seems, however, that sequential inference is also interesting from the point of view of general methodology, for it represents a natural model of the scientist's decision to stop accumulating (homogenous) evidence and choose one of the available hypotheses. The nature of this decision and its possible justification can be conveniently studied by means of this model.

It will be assumed throughout that the variables X_i are independent and identically distributed under each h_k. We adopt the notation

$$p(x \mid h_k) = \Pr\{X_i = x \mid h_k\}, \tag{1}$$

so that for $e_n = \langle x_1, \ldots, x_n \rangle$ we have

$$p(e_n \mid h_k) = \Pi_{i=1}^{n} p(x_i \mid h_k). \tag{2}$$

The last expression is the probability, under the assumption that h_k is true, of the sequence of outcomes $x_1, \ldots, x_n$. We will also call it the likelihood of h_k, given the evidence e_n. Prior probabilities of hypotheses h_k, if assumed to be known, will be denoted by $p(h_k)$.

Our discussion of sequential inference will begin by a rule defined in terms of likelihoods only (generalization of sequential probability ratio test). Then, assuming prior distribution, I will define a sequential rule based on average epistemic utility. The two approaches will be compared, under the assumption of prior equidistribution of hypotheses. Finally, I will interpret sequential inference as a search for information: the accumulation of evidence stops when a given level of information has been attained. The meaning of 'information' in this context may be either classical (i.e. based on entropy), or pragmatic (i.e. defined in terms of epistemic decision problem); the two approaches turn out to be closely related.

2 STRENGTHENED MAXIMUM LIKELIHOOD

Wald's work was based on the assumption, universally accepted at the time, that inductive inference should not presuppose the knowledge of prior probabilities of hypotheses. In accordance with this belief, his sequential rule of inference was defined in terms of likelihoods only. I will present it here in a generalized form (Szaniawski, 1961).

Let $[A_{rs}]$ be a $N \times N$ matrix, its entries satisfying $A_{rs} \geq 1$, $A_{rr} = 1$.
Rule R: Hypothesis h_r is accepted on evidence e_n if

$$p(e_n \mid h_r) \geq A_{rs} p(e_n \mid h_s) \quad \text{for all } s. \tag{3}$$

If there is no h_r satisfying (3), then no conclusion is drawn and the $(n + 1)$th experiment is performed.

The rule R is an obvious generalization of Wald's sequential probability ratio test: Wald's test is obtained by postulating $N = 2$. And it has analogous properties which justify its use.

Thus, if we denote by $\beta_{rs}(r \neq s)$ the probability of accepting h_r when in fact h_s is true, we have

$$\beta_{rs} \leq (1/A_{rs}) \quad \text{for all } s \neq r. \tag{4}$$

The proof of (4) is a straightforward generalization of Wald's argument. We will say that e_n is of the type r if the relation

$$p(e_m \mid h_t) \geq A_{ts}\, p(e_m \mid h_s) \quad \text{for all } s \neq t \tag{5}$$

is not satisfied for $m = 1, \ldots, n-1$, $1 \leq t \leq N$, and is satisfied for $m = n$, $t = r$. According to the rule R, hypothesis h_r is accepted on the evidence e_n iff e_n is of the type r. By definition, e_n of the type r satisfies condition (3), which means that the probability of obtaining e_n of the type r is at least A_{rs} times greater under the assumption that h_r than under the assumption that h_s.

Therefore, the probability measure of all $e_n(n = 1,2, \ldots)$ of the type r under the assumption that h_r is at least A_{rs} times their probability measure under the assumption that h_s. Formally,

$$\text{for all } s \neq r: \omega_r \geq A_{rs}\, \beta_{rs}, \tag{6}$$

where ω_r is the probability of accepting h_r when h_r is true. Since (6) implies (4), this completes the proof.

The inequality (4) defines an upper bound for the probability of error consisting in the acceptation of the hypothesis h_r when some other hypothesis h_s is true. Errors of that type may, in general, be of unequal importance, which will find expression in unequal values of the constants A_{rs}. However, the essential property of rule R, expressed by (4), is that the probability of *any* error can be made as small as desired by increasing the value of respective parameter of the rule. Of course, the average number of experiments needed to arrive at some conclusions would be thereby increased.

In analogy with any algorithm, there ought to be a guarantee that the procedure will terminate after a finite number of steps. Since the rule R is based on random events, postulated property of R must also be probabilistic. It can be described as follows. The probability of accepting some hypothesis after at most n experiments tends to 1 as n increases. Whether this is the case or not, depends, of course, on the probability distributions $p(x \mid h_k)$. For ar least some applications of R, the last mentioned property can be proved. Thus, for instance, let h_k be interpreted as stating that $\mu = \mu_k$, where μ is the mean of normal distribution with known variance. The proof (Szaniawski, 1961), that R leads to the acceptance of some h_k with the probability equal to 1 in the limit, rests essentially on the law of large numbers.

The rule R has $N(N-1)$ parameters, viz. the constants $A_{rs}(r \neq s)$. Such, at least, is the case when the inference is intended to serve a practical purpose, well defined in advance. In terms of this purpose, there is, in general, a preference for some errors over certain others. If, however, the inductive procedure is governed by a purely theoretical goal, there is no reason for treating the hypotheses in Z (hence, the errors) asymmetrically. The number of parameters can be then reduced to 1, by postulating

$$A_{rs} = A > 1 \quad \text{for } r \neq s. \tag{7}$$

The condition (3) takes on the form

$$p(e_n \mid h_r) \geq Ap(e_n \mid h_s) \quad \text{for all } s \neq r. \tag{SML}$$

It will be called 'strengthened maximum likelihood' condition, for rather obvious reasons. The maximum likelihood condition demands that this hypothesis is accepted which maximizes likelihood with respect to e_n. Allowing A in SML to be equal to 1, we obtain the statement that the likelihood of h_r maximum.

There is one important consequence of strengthening the maximum likelihood principle by means of $A > 1$: it may well happen that no h_r satisfies the SML condition. Hence, the rule of acceptance based on SML is necessarily sequential, whereas the maximum likelihood principle leads to a conclusion for any evidence and is, therefore, connected with fixed-size experimentation.

The rule R simplified by means of (7) will be called R_A. Thus, the rule R_A recommends the acceptance of h_r on the evidence e_n if h_r satisfies the SML condition.

3 MINIMUM OF EPISTEMIC UTILITY

If we allow the use of prior probabilities of hypotheses a different approach is possible. We now can calculate posterior probability distribution of hypotheses in Z, so that it makes sense to speak of the (average) utility of accepting h_r on the evidence e_n. An intuitively appealing sequential procedure would then consists in experimenting until one hypotheses, say h_r satisfies the following condition: the utility of accepting h_r exceeds a preassigned threshold value.

The choice of the utility function is determined by the nature of the present analysis. Since we are treating induction as a purely cognitive process, its goal can be defined as that of accepting the true conclusion. Let $u(h_r, h_s)$ be the epistemic utility of accepting h_r when h_s is true. We have

$$u(h_r, h_s) = \begin{cases} 1 & \text{for } r \neq s \\ 0 & \text{for } r \neq s. \end{cases} \tag{8}$$

The prior probabilities $p(h_k)$, together with likelihoods, determine posterior distribution $p(k_k \mid e_n)$:

$$p(h_k \mid e_n) = p(e_n \mid h_k)p(k_k)/C \quad C = \textstyle\sum_k p(e_n \mid h_k)p(h_k). \tag{9}$$

We average the utility function over all h_s, by means of $p(h_s \mid e_n)$. As a result we obtain $p(h_r \mid e_n)$ (cf. Marschak, 1974; Szaniawski, 1974). In other words, the epistemic utility of accepting h_r on the evidence e_n is the posterior probability of that hypothesis.

Let B, satisfying $0 \leq B \leq 1$, be the threshold value of $p(h_r \mid e_n)$, if h_r is to be accepted. This determines the following sequential rule of inference.

Rule S_B : Hypothesis h_r is accepted on the evidence e_n if

$$p(h_r \mid e_n) = \text{Max}_s\, p(h_s \mid e_n) \geq B. \tag{10}$$

If there is no h_r satisfying (10) no conclusion is drawn and the $(n + 1)$th experiment is performed.

Condition (10) is, of course, the conjunction of the following two statements.

$$p(e_n \mid h_r)p(h_r) \geq p(e_n \mid h_s)p(h_s) \quad \text{for all } s; \tag{11}$$

$$p(e_n \mid h_r)p(h_r) \geq B \textstyle\sum_s (p_n \mid h_s)p(h_s). \tag{12}$$

The first of the above conditions postulates maximalization of posteriors probability, hence also epistemic utility as defined by (8). Clearly, for any evidence e_n there exist a hypothesis h_r that satisfies (11). This condition is, therefore, not characteristic of sequential inference, for which (12) plays an essential role. One might say that (12) requires the best alternative to be *good enough*; if no hypothesis satisfies this requirement one is forced to increase the evidence until there is one that does.

It would be interesting to study the relation between the two approaches, i.e. between the rules R_A and S_B. The comparison cannot be carried out unless prior probabilities in S_B are fixed, since R_A was defined in terms of likelihoods only. The simplest way to do this is by assuming prior equidistribution:

$$p(h_k) = 1/N \text{ for all } k. \tag{13}$$

Conditions (11) and (12) become now

$$p(e_n \mid h_r) \geq p(e_n \mid h_s) \quad \text{for all } s; \tag{ML}$$

$$p(e_n \mid h_r) \geq B \textstyle\sum_s p(e_n \mid h_s), \tag{MU}$$

respectively. The first condition is that of maximum likelihood, the second will be called the minimum utility condition.

4 THE RELATION BETWEEN THE RULES R_A AND S_B

We are now in the position to compare the two rules of inference, by studying the relation between SML on the one hand, and the conjunction of ML and MU on the other. Obviously, SML implies ML. What is the sufficient condition for SML to imply MU?

The answer is not difficult to find if we add up all the inequalities in SML. We then have

$$(N - 1)p(e_n \mid h_p) \geq A \textstyle\sum_{s \neq n} p(e_n \mid h_s). \tag{14}$$

On the other hand, MU is equivalent to

$$(1 - B)p(e_n \mid h_r) \geq B \textstyle\sum_{s \neq n} p(e_n \mid h_s). \tag{15}$$

It follows that

$$\text{SML} \Rightarrow \text{MU} \quad \text{if } B \leq A/(A + N - 1). \tag{16}$$

The result can be put in words as follows. If a hypothesis h_r is accepted on evidence e_n according to the rule R_A then it is also accepted on evidence e_n according to the rule S_B, provided that B is at most equal to $A/(A + N - 1)$. The last expression is an increasing function of $\underline{A}$ and a decreasing one of N. This accords with intuition: when the number of equiprobable hypotheses increases it becomes more difficult for the posterior probability of any hypothesis to reach the threshold value, so that the upper bound for B must be correspondingly lowered if MU is to be satisfied by some h_r.

In order to see what ensures the reverse relation between R_A and S_B, let us notice that in the light of (15), $p(e_n \mid h_r)$ is at least $B/(1 - B)$ times the sum of all the remaining likelihoods. Hence,

$$\text{MU} \Rightarrow \text{SML} \quad \text{if } A \leq B/(1 - B). \tag{17}$$

Thus, if A satisfies the inequality in (17), the rule S_B is stronger than R_A, in the sense that the acceptance of some h_r according to S_B implies its acceptance according to R_A.

As a corollary, we get that if $B > \frac{1}{2}$ then MU implies ML, so that the two conditions defining S_B merge into one. Incidentally, since $A > 1$, $B > \frac{1}{2}$ is a necessary condition for the inequality in (17) to hold.

Assuming $N = 2$ we can combine (16) and (17) into

$$\text{if } N = 2 \text{ and } A = B/(1 - B) \text{ then SML} \Leftrightarrow \text{MU}. \tag{18}$$

It turns out that in the case of two hypotheses, equivalence of the rules R_A and S_B is possible; and in such a case only.

It will be recalled that both R_A and S_B had initially a more general form. The comparison was made possible by the simplifying assumptions (7) and (13) which make all the parameters in R equal to A, and all the prior probabilities in S_B equal to $1/N$.

5 SEQUENTIAL INFERENCE AS THE SEARCH FOR INFORMATION

In defining a sequential procedure, the crucial problem is, what shall be the criterion according to which the evidence is judged conclusive. Two possible answers to this question were discussed above. The first criterion was defined in terms of likelihoods and justified as putting upper bounds on the probabilities of errors. The second one interpreted the conclusiveness of evidence as making it possible to attain a given level of epistemic utility.

Still another approach, perhaps even more intuitive, would consist in adopting information as the required criterion. On this view, evidence is considered conclusive if it provides sufficient amount of information with respect to the initial problem (in our case, which of the hypotheses in the set Z is true?). A measure of information once defined, the arbitrariness, present in any form of induction, is reduced to the choice of the minimum value of information.

Let us first consider the Shannon measure. If we adopt the usual symbol for entropy, information[17] provided by e_n on Z is defined as

$$I(e_n, Z) = H(Z) - H(Z \mid e_n). \tag{19}$$

The absolute entropy of Z being constant with respect to e_n, we may dispense with it and make our rule of inference depend on the second term only. Thus, we postulate that the process of accumulating evidence continue until the conditional entropy of Z, given the evidence, falls below a certain level D; the most probable hypothesis is then chosen.

Rule T_D: Hypothesis h_r is accepted on the evidence e_n if the following conjunction holds:

$$p(h_r \mid e_n) = \text{Max}_s\, p(h_s \mid e_n) \tag{20}$$

$$H(Z \mid e_n) = -\textstyle\sum_s p(h_s \mid e_n) \log p(h_s \mid e_n) \leq D, \tag{21}$$

where D is a constant, satisfying $0 < D < \log N$. Experimentation continues if (21) does not hold.

One simple relation between T_D and S_B is given by the following assertion.

$$\text{If } D < -\log B, \text{ (21) implies (10).} \tag{22}$$

In order to see this, let us assume that (21) is satisfied for $D < -\log B$. Let h_r be the most probable hypothesis, in accordance with (20). We then have

$$-\log B \geq D \geq H(Z \mid e_n) \geq -\textstyle\sum_s p(h_r \mid e_n) = -\log p(h_r \mid e_n). \tag{23}$$

Hence, $B \leq p(h_r \mid e_n)$, which is precisely what (10) asserts. We thus see that, for $D \leq -\log B$, T_D is stronger than S_B, in a sense that a hypothesis accepted according to T_D is also accepted according to S_B.

If prior equiprobability of hypotheses is assumed, condition (20) becomes ML, i.e. the postulate of maximum likelihood for h_r, while (21) assumes the following form.

$$-(1/W) \textstyle\sum_s p(e_n \mid h_s) \log p(e_n \mid h_s) + \log W \leq D, \tag{ME}$$

where $W = \sum p(e_n \mid h_s)$. ME here stands for maximal entropy.

In view of (23), ME implies MU for $D \leq -\log B$, hence by (17) it also implies SML if $A \leq B/(1 - B)$. All the three rules are, therefore, closely related when the hypotheses are initially equiprobable. This last assumption has, in terms of (19), a natural interpretation: other things being equal, prior equidistribution maximizes information on Z, provided by e_n. If $H(Z \mid e_n)$ were replaced by $I(e_n, Z)$ in (21), then out of all possible prior distributions, equidistribution would be the one for which (21) expresses the strongest requirement concerning the evidence.

A sufficient condition for the rule S_B to be stronger than T_D is slightly more complex. To fix the ideas, let us interpret log as log*e*. The function $-x \log x$

reaches its maximum for $x = 1/e < \frac{1}{2}$. Therefore, $B > 1 - 1/e$ implies

$$-x \log x \leq B \log B \quad \text{for all } x \geq B,$$
$$-x \log x \leq (1 - B)\log(1 - B) \quad \text{for all } x \leq 1 - B. \tag{24}$$

On the other hand, (10) implies

$$p(h_s \mid e_n) \leq 1 - B \quad \text{for all } s \neq r. \tag{25}$$

It follows from the above that if $B > 1/e$ then (10) implies (21), for any

$$D \geq -B \log B - (N - 1)(1 - B)\log(1 - B). \tag{26}$$

Of course, the Shanon expression (19) is not the only existing explication of the concept of information concerning an exhaustive set Z of mutually exclusive hypotheses. An example of alternative approach is the concept of pragmatic information, relative to epistemic utility (cf. Szaniawski, 1974).

The idea is to calculate the difference between the maximum average utility when e_n is present and the corresponding maximum when it is not, i.e. when average utility is determined by prior distribution on Z. In the case we are interested in, the decision consists in choosing a hypothesis out of Z, and the epistemic utility of any such choice is given by (8). Pragmatic information of e_n concerning Z, say $C(e_n, Z)$, is then defined simply as

$$C(e_n, Z) = \text{Max}_s\, p(h_s \mid E_n) - \text{Max}_s\, p(h_s). \tag{27}$$

Since only the first term depends on e_n, the second one may be dropped in the definition of a sequential rule based on C (or it may be subtracted from the threshold value). We are then led to the condition (10), defining the rule S_B.

Thus, the analogies between the rules T_D and S_B are due to the similarity between the entropy-based concept of information and that of pragmatic information, when the last one is relativized to the epistemic problem, as defined by (8). The similarity was examined in some detail in Szaniawski (1974), for the case when both (19) and (27) are averaged over all $e_n \in E_n$. When this operation is performed, its results, of course, is a property of fixed-size evidence, viz. its average information concerning Z. The last mentioned quantity is non-negative, whereas both (19) and (27) can be negative for some e_n; the only exception is the case of prior equidistribution, relative to which any outcome e_n has some positive information.

6 CONCLUDING REMARKS

All the rules of inference discussed above had a common structure, characteristic of sequential inference in general. It can be described as follows.

If a hypothesis, say h_r, is to be accepted on the evidence e_n, a specified relation must hold between the two. Let it be symbolized by $W(e_n, h_r)$. As soon as e_n is such that W holds for some h_r, the procedure is terminated by accepting this particular h_r.

The relation W might consist in the fact that h_r is at least A times as probable on h_r than on any other hypothesis (SML), or in the probability of h_r given e_n exceeding a threshold value (MU), or it may represent some other requirement. The point is that it imposes a necessary condition on the evidence e_n. It follows that the rule of inference must provide for the case when this condition is not satisfied by e_n. Going on to e_{n+1} is the answer.

A non-sequential rule associates a conclusion with *all* possible premises of the type e_n. Or if it allows, for some e_n, the 'no conclusion' verdict, no further steps are dictated by the rule. It follows that the number n of experiments can be specified in advance, since it depends exclusively on the subject's decision. The decision may be based on essentially the same consideration as those that govern the choice of the exact form of W. Let us, for instance, assume that the purpose of inference is adequately described by the function which associates with each e_n $\text{Max}_s\ p(h_s \mid e_n)$, coupled with the probability distribution of e_n. How the choice of n is influenced by the shape of the two functions, is something I am not going to discuss here. Certainly, there is no unique solution to this problem.

In the case of sequential inference, the subject must decide on the actual value of at least one parameter in his rule. If, for example, his purpose is given by (8) he has to choose the threshold value B. Now the number of experiments needed to reach a conclusion becomes a random variable and the consequences of the subject's choice are represented by the probability distribution of that variable. The situation is, in a sense, reverse to the one described above when n was fixed and the degree of the goal's achievement varied randomly.

Although the cost of experimenting has, so far, been left out of account, there is the implicit assumption that it exists and puts restrictions on the parameters of the inductive procedure. Otherwise, there would be no limit to the process of gathering evidence. If the cost of experimentation is explicitly introduced (and the purpose of local induction defined in the same units) a kind of balance can, in principle, be established, leading to a rule which maximizes the average net gain. I shall not pursue the topic further, since it seems hardly possible to inquire into the nature of this balance without making specific assumption between the hypotheses and the data.

To conclude the present discussion, a few words on the so-called 'truncated' rules. There is always a limit to the gain in utility, associated with a correct choice of hypothesis. Now, a sequential rule of inference provides, in general, no guarantee that the procedure will terminate before this limit is attained by the increasing cost of experimentation. Such a possibility cannot be excluded, even if it is highly improbable. Clearly, to continue experiments beyond the above mentioned limit would be inconsistent with the meaning of utility. Hence it is necessary to modify the sequential rule by adding a proviso which at a certain point puts a stop to experimentation, even if the evidence does not satisfy the condition W that defines the rule. The use of 'truncated' rules can, therefore, be justified on purely theoretical grounds if both the purpose of inference and the limited character of data are explicitly stated.

NOTES

[1] Actually, Wald's work dates from 1943. For the duration of the war the results were classified, which delayed their publication.

[2] Hintikka (1968) calls it 'transmitted information', averaged over the elements of Z.

REFERENCES

Hintikka, J., 1968, 'The Varieties of Information and Scientific Explication', in B. van Rootselaar and J. F. Staal (eds.), *Logic, Methodology and the Philosophy of Science* III, North-Holland Publishing Company, Amsterdam, 1968, pp. 311–332.

Marschak, J., 'Information, Decision and the Scientist', in C. Cherry (ed.), *Pragmatic Aspects of Human Communication*, Theory and Decision Library, Vol. 4, D. Reidel Publishing Company, Dordrecht and Boston, 1974, pp. 145–178.

Szaniawski, K., 1961, 'A Method of Deciding Between *N* Statistical Hypotheses', *Studia Logica* 12, 135–143 [see this volume, pp. 80–86.].

Szaniawski, K., 1974, 'Two Concepts of Information', *Theory and Decision* 5, 9–21 [see this volume, pp. 157–166.].

Wald, A., 1947, *Sequential Analysis*, J. Wiley, New York.

13

INTERPRETATIONS OF THE MAXIMUM LIKELIHOOD PRINCIPLE*

0 INTRODUCTORY REMARKS

The maximum likelihood principle is one of the simplest and most intuitive rules of non-deductive inference. It was originally formulated by Ronald A. Fisher [2] for the use of mathematical statistics, and more particularly for its part concerned with point estimation, i.e. with determining the numerical value of parameter, for example, a mean. But the idea expressed by the maximum likelihood principle has a more general character. It may be formulated as follows.

Let H be an n-element set of sentences, the exclusive disjunction of which is true. We shall call *hypotheses* the elements of the set H.

Let e be a sentence the truth of which has been established. The sentence e (evidence) will be called the *premise* or *empirical data.*

Moreover, let there be for each $h \in H$ a given $p(e \mid h)$, i.e. the probability of e under the assumption that h. The chosen notation would indicate that it is a conditional probability. Indeed, the expression $p(e \mid h)$ can be interpreted in this way. But in the case when the meaningfulness of assigning probability to hypotheses raises doubts, this expression will be considered as the probability of the sentence e, parametrical with respect to h. It is anyway substantial that e is constant while h ranges over the set H.

The question answered by the maximum likelihood principle is, of course, that of the choice of hypothesis in the light of empirical data. As it is known, the answer is: the hypothesis should be chosen which maximalizes $p(e \mid h)$.

In other words, the maximum likelihood principle tells us to choose the hypothesis with relation to which the empirical data are the most probable. The name of the rule originates from the fact that $p(e \mid h)$, i.e. the probability of e with respect to h, was called by Fisher the likelihood of the hypothesis h with relation to the data e. Thus the considered rule promotes the choice of the hypothesis most likely in the light of empirical data.

This view of the likelihood of a hypothesis is, of course, something different from its probability. The origin of the maximum likelihood principle amount precisely to that: it was meant to formulate an inference rule which would not refer to probabilities of hypotheses. These probabilities – if they are to be interpreted objectively – are rarely given, and hence it would be unrealistic to make an inference rule dependent on them. Contemporary mathematical statistics is wholly based on this view. It was only a more common use of the concept of subjective probability that made it possible in the last few decades to develop the so-called Bayesian theory of probabilistic inferences, in which the maximum likelihood principle and related rules do not play any substantial role (cf. Szaniawski [7]).

It is not difficult to indicate the reason for which the maximum likelihood principle seems intuitively convincing. Namely, it is a generalization of the commonly used procedure of elimination of hypotheses: the hypothesis is rejected in the light of which the experimentally confirmed sentence e is minimally probable. This reasoning, which Czerwiński [1] proposed to call 'weakened *modus tollens*', is somehow analogous to its deductive counterpart. Instead of the implication it contains a high conditional probability of the negation of the sentence e, and as the truth of e has been established, it is the basis for the rejection of the antecedent of this quasi-implication. By repeatedly applying this elimination procedure to the hypotheses of the set H, we gradually reduce that set, and eventually the only hypothesis (hypotheses) left is (are) the one (those) for which $p(e \mid h)$ assumes the maximum value – conformingly to the maximum likelihood principle.

It naturally remains an open question whether the above-presented justification should be considered satisfactory. No matter, however, how this question is resolved, it is no doubt advisable to realize the possible interpretations of the maximum likelihood principle in the categories of the aim which directs the choice of a hypothesis. Such an aim should be specified in terms of a function evaluating the possible result of inference (and called, following C. Hempel, epistemic utility). Then it should be found out in what circumstances the maximum likelihood principle leads to the maximalization of that function.

The remarks presented below concern the problem formulated as above. They are partially a repetition of already known views and partially their extension.

1 MAXIMALIZATION OF PROBABILITY OF CONCLUSION

Probably the simplest aim of an inference is to accept a true hypothesis. If the task is reduced to this, the evaluative function – let us call it u_1 – gives the same value to all correct conclusions as well as to all erroneous cases. It only differentiates error and non-error, to the advantage of the latter. This leads to the following epistemic utility:

$$u_1(h, k) = \begin{cases} 1 & \text{if } h = k, \\ 0 & \text{if } h \neq k. \end{cases} \tag{1}$$

It is the function of two variables: the accepted hypothesis and the true one; i.e. both h and k range over the set H. In the case of identity of arguments the conclusion is true; in the opposite case one of the $n(n-1)$ possible errors occurs.

The mean value of $u_1(h, k)$, naturally equal to the probability of h, is the epistemic utility of accepting the hypothesis h. Hence the aim determined by (1) amounts to the maximalization of that probability which, according to Bayes' theorem, is given by the formula:

$$p(h \mid e) = p(e \mid h) \cdot p(h) \cdot C, \tag{2}$$

where

$$C = [\textstyle\sum_h p(e \mid h)p(h)]^{-1} \tag{3}$$

is a constant factor normalizing the probability.

It immediately results from (2) that if for all h

$$p(h) = 1/n, \tag{4}$$

i.e. all hypotheses are *a priori* equally probable, then the maximalization of $p(h \mid e)$ is equivalent to the maximalization of the likelihood. The maximum likelihood principle is therefore optimal with respect to the aim determined by u_1 – under the assumption of equiprobability of hypotheses.

Let us remind that the so-called principle of insufficient reason, attributed to Laplace,[1] tells us to consider hypotheses equally probable if nothing indicates that one of them should be distinguished. As it can be seen, by using the principle of insufficient reason likelihood is maximalized.

2 MAXIMALIZATION OF TRANSMITTED INFORMATION

Another interpretation of the likelihood principle has been given by Hintikka [3]. It is based on the concept of transmitted information. Taking as a starting point the known measure of information contained in a sentence s:

$$\mathrm{inf}(s) = -\log p(s), \tag{5}$$

Hintikka defines the information transmitted by the sentence h about the subject matter of e as follows:

$$\text{trans inf}(h \mid e) = \mathrm{inf}(e) - \mathrm{inf}(e \mid h) = \log[p(e \mid h)/p(e)]. \tag{6}$$

Here is the justification of this definition: "inf($e \mid h$) is the information e adds to that of h, i.e. the uncertainty that there remains concerning e even after we have learned that h is true. Hence inf(e)-inf($e \mid h$) measures the reduction of our uncertainty concerning e which takes place when we come to know, not e, but h." (Hintikka [3], p. 316.)

To explain the sentence e, Hintikka says, means to indicate an h such that it will transmit possibly as much information as it is stated by e. "What we want to do is to find an h such that the truth of e is not unexpected, given h" (p. 321). If, therefore, the aim to be achieved by choosing h is the explanation of e, then the corresponding epistemic utility, let us say u_2, is as follows:

$$u_2(h) = \log[p(e \mid h)/p(e)]. \tag{7}$$

The maximalization of u_2 is therefore equivalent to the maximalization of $p(e \mid h)$, and this precisely is the maximum likelihood principle.

"Thus we arrive at the famous maximum likelihood principle as the natural method of choosing one's explanatory hypothesis in the kind of circumstances indicated (. . .). Thus the importance of this principle in statistics has an interesting general reason which can be brought out in terms of the concept of information" (p.322).

The peculiarity of this approach lies in that it does not take into consideration the logical value of h, i.e. of the conclusion, at all. The epistemic utility is here the function of one argument only, i.e. of the accepted hypothesis, and it does not depend on which of the hypotheses $h \in H$ is true. I believe this can be explained in two ways.

The first explanation requires a change of assumptions. Let us suppose that the hypotheses of the set H are not pairwise exclusive, but conversely: that they have all been found true. The task is, then, to find the best explanation of e among those true sentences. Such a procedure may be considered, although for obvious reasons it could be called an inference.

Another way of understanding Hintikka's proposed maximalization of u_2 consist in considering it as the first stage of the explanation of e. It would consist in distinguishing in the set H of hypothesis suitable to explain e – in the sense that it conveys most information about e. The hypothesis thus distinguished could then be accepted as an explanation – provided its truth was somehow demonstrated, or at least made probable. Such a two-stage explanatory procedure could hardly be objected to (the first stage would consist in utilizing the maximum likelihood principle), but it should be realized that the status of the discussed principle would then undergo a basic modification. For, according to the classical approach, the maximum likelihood principle is a rule of inference; its application therefore leads to the acceptance of the sentence h as true, h being correlated as a conclusion to the sentence e, i.e. the premise. But the principle, proposed by Hintikka, of distinguishing an element of the set H by maximalizing u_2 does not have this property. For there is no justification of a procedure which consists in accepting as true hypothesis h maximalizing u_2 – because the sentence e has been accepted as true. If we were to follow Hintikka and accept that the maximum likelihood principle is justified by its function of explaining an established fact, then the very same principle could no longer be considered a rule of inference – contrary to the original intention of its author and probably contrary to the practice of its applications.

3 MAXIMALIZATION OF THE PROBABILITY OF CONCLUSION AND OF TRANSMITTED INFORMATION

From the above considerations it follows that it is a necessary condition for the maximum likelihood principle to maintain its status of a rule of inference according to which the epistemic utility maximalized by that principle allows to evaluate and differentiate truth and falsehood. In other words, epistemic utility should fulfil the following condition (for all h, k such that $h, k \in H$, $h \neq k$):

$$u(h, k) > u(k, h). \tag{8}$$

If, moreover, Hintikka's idea is to be preserved (i.e. the evaluation is to take into consideration the degree to which the premise is explained by the hypothesis), then the function u should fulfil two further considerations, which we shall formulate as follows:

For all h, k and l non-identical in pairs, of the set H:

$$\text{trans inf}(h \mid e) > \text{trans inf}(k \mid e) \tag{9}$$

entails

$$u(h, h) > u(k, k) \tag{10}$$

and

$$u(h, l) > u(k, l). \tag{11}$$

The condition (9) ⇒ (10) postulates the preferential ordering of true conclusions with respect to trans inf. The condition (9) ⇒ (11) postulates the same for false conclusions.[2]

There is of course more than one function u fulfilling the above conditions. Within the limits determined by these conditions the choice of a concrete epistemic utility should most likely follow the criterion of simplicity. One of the simplest is the device proposed in Szaniawski [8].

Let us suppose that t is a function which is to generate preferential orderings. Let it be a positive function. Then the fulfilment of the condition (8) may be assured by defining u as follows:

$$u(h, k) = \begin{cases} t(h) & \text{for } h = k, \\ t(h) - \operatorname{Sup} t(h) & \text{for } h \neq k. \end{cases} \tag{12}$$

It is clear that the condition (8) is fulfilled as $t(h) - \operatorname{Sup} t(h) < 0 < t(h)$. The fulfilment of the other two conditions is an immediate consequence of the definition (12).

The identification of $t(h)$ with trans inf $(h \mid e)$ is hindered only by the fact that the latter function may assume negative values. This is not, however, a substantial obstacle. It suffices to note that it is a sufficient condition for the positivity of trans inf $(h \mid e)$ that

$$p(e \mid h) > p(e). \tag{13}$$

This condition states the positive probabilistic correlation between the hypothesis h and the sentence e. But only such hypotheses can pretend to the role of explanation of the sentence e. Therefore, if a certain h does not fulfil the condition (13), it can be eliminated from the set H (of course, only if the explanation of e codetermines the choice of a hypothesis). This entails that the assumption of the positivity of trans inf is in this case no substantial limitation.

Thus, substituting in (12) trans inf $(h \mid e)$ for $t(h)$ and considering

$$\text{Sup trans inf}(h \mid e) = \inf(e) = -\log p(e) \tag{14}$$

we obtain the epistemic utility u_3 fulfilling the above-formulated conditions for u:

$$u_3(h, k) = \begin{cases} \log[p(e \mid h)/p(e)] & \text{if } h = k, \\ \log p(e \mid h) & \text{if } h \neq k. \end{cases} \tag{15}$$

Determining the mean of k in terms of the probability distribution function $p(k \mid e)$, we obtain the average utility connected with the acceptance of the hypothesis h:

$$E_{p(k \mid e)} u_3(h, k) = \log p(e \mid h) - p(h \mid e) \log p(e). \tag{16}$$

We can now answer the question concerning the assumption under which the maximum likelihood principle is equivalent to the maximalization of transmitted information, if that maximalization is limited by the condition (8). Because of (16), the answer is the same as when the only aim was to derive a true conclusion: the maximum likelihood principle may be interpreted as the maximalization of u_3 if and only if the hypotheses of H are *a priori* equiprobable.

4 SEQUENTIAL GENERALIZATION

It is worth remarking that the maximum likelihood principle is a special case of a procedure which assumes the possibility of accumulating empirical data. It is so, for example, when we successively examine the elements of a set (not necessarily finite), in order to decide at some moment that the accumulated data are sufficient to chose one of the hypotheses from that set. Let us denote the data accumulated after n stages of research by e_n; of course, for every n: $e_{n+1} \Rightarrow e_n$.

A sequential rule of inference (Wald [9]) is one which takes into consideration the character of the data outlined above: for every n it determines e_n for which one of the hypotheses (and which) is to be accepted, and e_n for which no choice could be made and one should then proceed to e_{n+1}. There is a sequential rule of inference (Szaniawski [5]) let us call it R – which is a generalization of the maximum likelihood principle.

The Rule R. Each pair of different hypotheses h, k of the set H has been assigned a certain constant $A_{hk} \geq 1$. At the nth stage of research the hypothesis h should be accepted if and only if

$$\text{for every } k \neq h\colon p(e_n \mid h) \geq A_{hk} \cdot p(e_n \mid k); \tag{17}$$

in the case when (17) does not hold for any h, proceed to e_{n+1}.

It is easy to note that rule R becomes the maximum likelihood principle if we assume that

$$A_{hk} = 1 \quad \text{for all } h, k \in H. \tag{18}$$

The condition (18) eliminates the sequential character of rule R: since the set H is finite, there is at last one[3] hypothesis $h \in H$ fulfilling (17) for $A_{hk} = 1$, i.e. proceeding to e_{n+1} is unnecessary. Thus the generalization of maximum likelihood principle in the form of rule R is natural. It preserves what is substantial for principle: the comparison of hypotheses in terms of their likelihood. But some elasticity is introduced into this comparison. Namely, the inference maker is given the possibility of non-symmetrical treatment of hypotheses – by an appropriate choice of the constants A_{hk}. If, for example, this person is particularly interested in avoiding the rejection of a distinguished hypothesis h when it is true, he will choose a relatively small A_{hk}, for all k. Different choices of numerical values of the parameters of rule R would be directed by analogical reasons. The assignment of values different from 1 to these parameters entails the sequential character of the rule R. For it is not generally excluded that no hypothesis h fulfils the condition (17), which requires to suspend the opinion and to extend the empirical basis.

A guarantee is of course necessary that this process will terminate – with a probability equal to one – in a finite number of steps, leading to the acceptance of one of the hypotheses of the set H. It is possible to demonstrate this property of the rule R only by assuming a more exact characterization of the nature of the probabilistic relation between the members of H and the successive extensions of the empirical data. In the paper mentioned above (Szaniawski [5]) the demonstration has been carried out for two rather important cases.

A similar precision would also be required for the examination of the rule R from the point of view adopted in the sections 1–3 of the present paper. This would naturally be a more complicated question.

5 SUMMARY

The above remarks addressed the following problem: under what assumptions is it possible to interpret the maximum likelihood principle as the maximalization of the aim of research understood as: (1) acceptance of the true conclusion, (2) acceptance of the best explanation of the empirical data. With respect to the first aim, the maximum likelihood principle is optimal under the assumption that all the hypotheses are *a priori* equiprobable. With respect to the second aim (understood in agreement with Hintikka's proposal) it is optimal without any additional assumptions – but only on the condition it does not permit to accept the distinguished sentence as true, i.e. – that it is not a rule of inference. If the aims (1) and (2) are put together the optimality of the maximum likelihood principle is again conditioned by the *a priori* equiprobability of the hypotheses of the set H.

In the final section the maximum likelihood principle has been presented as a special case of a (sequential) rule of inference.

NOTES

* First published in *Rozprawy Logiczne*, Warszawa 1964. Translated by S. Wojnicki.

[1] Hence the so-called Laplace's criterion in decision-making. The paper by Szaniawski [6] has demonstrated that under certain assumptions the MLP is the optimal – in Laplace's sense – method of point estimation.

[2] The postulates adopted here are analogical to those formulated by J. Levi for other purposes (cf. Levi [4], p. 76).

[3] It is possible (which has already been mentioned) that more than one hypothesis fulfil (17), i.e. maximalize $p(e_n \mid h)$. The method of dealing with non-uniqueness, for example, by an arbitrary choice of the hypothesis, is not relevant here.

REFERENCES

[1] Z. Czerwiński, 'On the Relation of Statistical Inference to Traditional Induction and Deduction', *Studia Logica* 9 (1958) 243–264.

[2] R. A. Fisher, 'On the Mathematical Foundations of Theoretical Statistics', *Philosophical Transactions of the Royal Society* 222, *Series A* (1922).

[3] J. Hintikka,'The Varieties of Information and Scientific Explanation'. In: *Logic, Methodology and Philosophy of Science III* (ed. by Rootselaar and Staal), North Holland (1968) 311–331.

[4] J. Levi, *Gambling with Truth. An Essay on Induction and the Aims of Science*, Alfred A. Knopf, New York 1967.

[5] K. Szaniawski, 'A Method of Deciding Between *N* Statistical Hypotheses', *Studia Logica* 12 (1961) 135–143.

[6] K. Szaniawski, 'Zasada największej wiarygodności (próba częściowej oceny)', in: *Rozprawy Logiczne. Księga pamiątkowa ku czci Kazimierza Ajdukiewicza* ['The Maximum Likelihood Principle – an Attempt at Partial Evaluation', in: *Logical Dissertations. Papers in the Honour of Kazimierz Ajdukiewicz*] PWN, Warszawa 1964 [see this volume, pp. 106–113.].

[7] K. Szaniawski, 'Dwie koncepcje indukcji', in: *Fragmenty Filozoficzne, seria trzecia. Księga Pamitkowa ku czci Tadeusza Kotarbińskiego* ('Two Concepts of Induction', in: *Philosophical Fragments, Third Series. Papers in Honour of Tadeusz Kotarbiński*), PWN, Warszawa 1967.

[8] K. Szaniawski, 'Types of Information and Their Role in the Methodology of Science', *Proceedings of the Conference on Formal Methods in Methodology of Science*, Jabłonna 1974 [see this volume, pp. 167–176.].

[9] A. Wald, *Sequential Analysis*, J. Wiley, New York 1947.

14

SOME REMARKS CONCERNING THE CRITERION OF RATIONAL DECISION-MAKING

The purpose of this paper is to discuss some aspects of so-called 'decision problem under uncertainty' (d.p.u.u.) which could be roughly described as follows.

Suppose that a person P is faced with the problem of choosing a course of action from an exhaustive set a of alternative actions available: $A_1, A_2, \ldots, A_m$. And suppose further that $S = \{S_1, S_2, \ldots, S_n\}$ is an exhaustive set of possible 'states of Nature' relevant to the choice in the following sense: the desirability (for P) of any action form the set a depends upon the state of Nature (from the set S) which actually occurs. To put it more precisely: if an action A_i is chosen and the state S_j occurs, then the situation (A_i, S_j) has for the decision-maker P a specified utility[1] u_{ij}. The purpose of P in making the choice is to obtain as large utility as possible.

Now, if the decision maker P *knows* which state of Nature occurs, the solution is trivial. If the probability distribution over set S is known to him, it is again not difficult to say what would be the rational way for him to behave: 'choose such an action A_i for which the expected utility is maximum'.

But suppose no information is given concerning the chances of S_j to happen. All the decision-maker knows are possible consequences of his choice. If this is the case, which ones of all possible choices are rational? This is to ask for a *criterion of choice*; having adopted a criterion we might say that every actual choice made according to that criterion is *ex definitione* a rational one. But then again the choice of criterion must be justified somehow, although it is suspected that this justification could not possibly be entirely convincing for everybody: the data of the problem are too weak to allow such a solution which would conform to all the different (and rather vague) connotations of the word 'rational'.

It seems, therefore, that the only possible way to proceed is to examine properties of several criteria of choice and, having decided which properties are to be considered basic for the word 'rational', choose this criterion which possesses them. Or, to put it the other way round: to formulate set of postulates (or necessary conditions) for the criteria of choice is to fulfil and adopt the criterion which fulfils the greatest number (or the most important) of them all.

This means that every criterion of choice constitutes possible clarification (an explication) of the meaning of the expression 'rational way of behaviour' in the above described situation of choice. This is what makes the problem extremely interesting from the philosophical point of view. It might even be said that such a problem (how to explicate the expression 'rational way of making a decision' in the given context) comes under a heading of praxiology[2] which is a philosophical discipline. The methods of solving it are, however, to a large extent mathematical.

Against the view that the d.p.u.u. has considerable philosophical importance there is the objection that such an explication of the term 'rational choice' can be applied (in the strict sense) only to a relatively small number of cases, as the assumption that all the relevant states of the Nature are known and the utility matrix given is rather a strong one. It has, indeed, been asserted by some writers that in the d.p.u.u. a highly artificial situation is described; that in actual situations of choice either the utility matrix is not given or the ignorance is not total, i.e. some partial information about the relative likelihood of the states of Nature is almost always at hand.[3] It may be argued, however, that even if d.p.u.u. proves to be an idealized model of actual problems of choice, this fact alone does not deprive it of considerable importance. To construct such models is certainly not a bad way to proceed in describing real life situations. And anyway, this is only the first step which can perhaps be improved upon, e.g., following the line indicated by Simon. There is also another point which adds to the importance (theoretical at least) of d.p.u.u. Suppose that some empirical evidence can be obtained concerning the likelihood of the states of Nature to occur; and that this evidence is of the kind very often met in science, i.e., every possible outcome x of an experiment has a certain probability, say $p(x/S_j)$, under the assumption that S_j occurs. If the decision-maker wants to take into account such experimental evidence, then the problem is no longer a simple one: which action to choose, but rather: how to associate the choice of action with the outcome of the experiment. Any function $d(x) = A_i$ provides a possible answer to the last question and with every pair (d, S_j) an expected utility can be associated. Thus the problem of making decisions in the light of experimental evidence (of a certain kind) essentially reduces to d.p.u.u.[4]

It is perhaps worth mentioning that the utility function has been introduced in a sufficiently general way to express preferences of many kinds. Thus a d.p.u.u. may describe a situation of choice in the face of possible monetary losses and gains, and of otherwise (e.g. ethically) evaluated outcomes as well. This obviously makes the field of applications rather large. Some other interpretations of d.p.u.u. are also known to exist,[5] e.g.,

1. Let a be a set of objects the person P has to choose from; the objects are differently valued from several points of view: $S_1, \ldots S_n$; u_{ij} then represents

an evaluation of the object A_i under its j-th aspect.

2. Let a be a set of alternative decisions differently valued by the individuals $S_1, \dots S_n$; u_{ij} then represents the evaluation of the decision A_i by the individual S_j and the problem is: which decision is to be adopted by the group $S = \{S_1, \dots S_n\}$. (This, of course, presupposes the possibility of interpersonal comparison of preferences – a very debatable assumption.)

The purpose of the above remarks was to stress the significance of the problem under discussion. And now, to come to the point.

It seems necessary, in the first place, to distinguish between the situations of repetitive and non-repetitive choice. By the non-repetitive choice situation we mean such a d.p.u.u. where the choice of action is to be made only once. In repetitive situation the decision-maker has to choose from the set a several times; if this is the case, the problem arises if it would not pay 'in the long run' to make the choice depend upon some randomizing device. The answer to the last question is affirmative and the notion of 'mixed strategy' (ascribing to every A_i a probability $\xi(A_i)$ of its being chosen) has been introduced accordingly.

What I want to stress here is that – whatever the criterion of choice adopted – the use of mixed strategy in a non-repetitive situation is not profitable in the sense of this criterion. The reason for this assertion is rather simple one: if an action A_i is the best in the sense of the criterion adopted, then the action A_j chosen in any random way is at the most as good as A_i (if $i = j$), and in general worse than A_i (if $i \neq j$). In what is going to follow we shall concern ourselves principally with non-repetitive situations; mixed strategies will, as a rule, not be considered.

Before we proceed to discuss existing criteria and postulates characterizing them, it will be necessary to introduce the concept of (weak) domination. The action A_k is said to be (weakly) dominated by A_l if for every j: $u_{kj} \leq u_{lj}$ and for some j: $u_{kj} < u_{lj}$. A well-known principle concerning the choice of action is based on this notion, usually called 'the non-domination principle'; it prohibits the choice of such an action A_k which is dominated by some other action A_l. This principle has the merit of not being subject to any doubt: on the other hand, it is rather weak, as it only rejects certain actions as 'ineligible'. After the elimination of dominated actions is performed, a (usually large) class of actions is still left to choose from; as a rule, such elimination does not reduce the number of possible choice in a drastic way.

It ought to be noted that even the process of reducing the class of all actions available to the (minimal) class of non-dominated actions might become very complicated if the matrix $\|u_{ij}\|$ is infinite; this is the case in almost all applications to the so-called statistical inference. It appears, therefore, that we have simplified considerably the problem by assuming both the set of actions and set of states of Nature to be finite. However, if our principal aim is to clarify the meaning of the expression 'rational choice', it is perhaps permissible to start from such a simple d.p.u.u. The problem then is: how to choose from a set of actions, none of which is dominated by any other. The usually suggested alternative answer to that question are the following.

1. The Laplacean criterion.

Choose such an action A_s which maximizes $\frac{1}{n}\sum_j u_{ij}$.

$$\frac{1}{n}\sum_j u_{sj} = \max_i \frac{1}{n}\sum_j u_{ij}.$$

2. The maximin criterion.
Choose such an action A_s which maximizes $\min_j u_{ij}$

$$\min_j u_{sj} = \max_i \min_j u_{ij}.$$

3. The Hurwicz family of criteria.
Choose such an action A_s which maximizes $\alpha \min_j u_{ij} + (1-\alpha)\max_j u_{ij}$, where the parameter α (called sometimes 'the pessimism index') fulfils: $0 \leq \alpha \leq 1$.

$$\alpha \min_j u_{ij} + (1-\alpha)\max_j u_{ij} = \max_i[\alpha \min_j u_{ij} + (1-\alpha)\max_j u_{ij}]$$

4. The minimax regret (Savage's) criterion.
Choose such an action A_s which minimizes $\max_j(\max_k u_{kj} - u_{ij})$.

$$\max_j(\max_k u_{kj} - u_{sj}) = \min_i \max_j(\max_k u_{kj} - u_{ij}).$$

The above criteria have this property in common that they associate with every action A_i a number which may be called 'the desirability of A_j according to the given criterion'; then the action with highest index is to be adopted. (In the case of Savage's criterion it is enough to change the sign of the expression within the brackets; the procedure is then equivalent to maximizing $\min_j(u_{ij} - \max_k u_{kj})$.) They thus go farther then the actual problem of choice: they establish a complete ordering of actions and, what is more, they even make it possible to compare their desirability.

The index in question is obviously a non-decreasing function of the utility entries for a given action, and in one case (Savage's criterion) it is even a function of all the entries in the utility matrix. We consider it a rather important feature of Laplacean criterion that it is the only one among the above-mentioned, for which the 'desirability index' of A_i is strictly increasing function of $u_{i1}, \ldots u_{in}$. Maximin takes into account only the smallest value of utility for a given action, and Hurwicz's concentrates on the smallest and biggest utility, leaving out all the intermediate values; some rather paradoxical choices recommended by these criteria are due to this fact.

The strong and weak points of the four criteria can be discussed in an 'informal' way. Such a discussion (an excellent one) can be found in the already quoted book by Luce and Raiffa, I shall not, therefore, repeat their arguments. A more formal approach is by way of 'postulates for rational choice' which provide means of appraising the criteria. Following this line, I shall adopt as a starting-point the set of axioms given by Luce and Raiffa.[6] I shall have to reproduce them here, using the notation adopted by the authors.

Let (α, S, u) be a d.p.u.u; '$\hat{\alpha}$' will denote the set of actions chosen; $\hat{\alpha}$ is a subset of α (and will be called the choice-set), '$\sim$' will denote equivalence between two actions (i.e. their having the same utility for every state of Nature). '$>$' will denote (weak) domination. An action A not dominated by any other action in α will be called admissible.

The suggested postulates for the solution $\hat{\alpha}$ of every d.p.u.u. are:

1. $\hat{\alpha} \neq 0$, i.e. every problem can be solved.
2. $\hat{\alpha}$ is invariant under a transformation $au + b$; $(a > 0)$ of the utility function (this follows from the axioms of utility).
3. $\hat{\alpha}$ is invariant under the re-labeling of actions.
4. If $A \in \hat{\alpha}$ and $A' > A$ or $A' \sim A$, then $A' \in \hat{\alpha}$.
5. If $A \in \hat{\alpha}$, then A is admissible[7]
6. $\hat{\alpha}$ is invariant under adding to α a new action dominated by or equivalent to some old action.
7. An action not belonging to $\hat{\alpha}$ cannot be made to belong to $\hat{\alpha}$ by adding new actions to α.
8. $\hat{\alpha}$ is invariant under adding a constant to each entry in the column.
9. If $A \in \hat{\alpha}$ and $A' \in \hat{\alpha}$ then each probability mixture of A and A' belongs to $\hat{\alpha}$. (By 'probability mixture' is meant a procedure consisting in choosing A with probability η and A' with probability $1 - \eta$.
10. $\hat{\alpha}$ is invariant under re-labeling of states of Nature.
11. $\hat{\alpha}$ is invariant under deleting repetitious column.

It is known that the following axioms[8] are not satisfied by some of the criteria: 11 by the Laplacean criterion; 8 by maximin and Hurwicz criteria; 9 by the Hurwicz criteria for $\alpha \neq 1$; 7 by minimax regret criterion.

It seems that not all the axioms violated by the criteria are equal importance. We consider the case of axioms 9 and 7 to be the most serious one.

Concerning the axiom 9, postulating the convexity of the set $\hat{\alpha}$, it might be said that if two actions A and A', are considered optimal *in any sense*, then it seems highly paradoxical to rule out a procedure leading ultimately to the choice of A and A'. And the probability-mixing of A and A' is precisely such a procedure. This is valid in spite of the fact that we have ruled out the use of mixed strategies: choosing in a random way from a set $\hat{\alpha}$ of the acts considered optimal (in the sense of some criterion) is certainly something very different from introducing a new action, being a probability mixture of *all the actions available.* It thus seems that the position of Hurwicz criterion (in all the cases when it does not reduce to maximin) is considerably weakened by its not satisfying the axiom 9.

As regards axiom 7, its reasonableness is rather obvious. If an action A does not belong to the choice-set $\hat{\alpha}$, this means that there is in $\hat{\alpha}$ at least one action which is better than A in the sense of the criterion adopted (because the set $\hat{\alpha}$ is assumed by axiom 1 to be non-empty). It seems hardly rational to assume that this fact could be changed by adding any new action to the set α. It is, therefore, rather difficult to call rational a choice made according to a principle that violates (as does the minimax regret criterion) axiom 7.

Having argued for the postulates 7 and 9, not satisfied respectively by minimax regret and Hurwicz criteria, I am now going to state briefly the reason for rejecting axioms 8 and 11. As to the axiom 8, it has already been pointed out by some authors that it is incompatible with any degree of what one might call 'cautiousness'. To speak in terms of an example, let as consider two simple problems of choice:

I

	S_1	S_2
A_1	0	c
A_1	c	0

II

	S_1	S_2
A'_1	0	0
A'_2	c	$-c$

$c > 0$

It is a fair guess that every decision-maker would be indifferent between the actions A_1 and A_2 in the first problem, since there seems to be no reason to prefer one of them to the other. But not everybody would be indifferent between A'_1 and A'_2 in the second problem, since in choosing A'_2 one incurs the risk of losing c utilities (although 'balanced', it is true, by the possible gain of c) while in choosing A'_1 one incurs no such risk at all. And yet, according to axiom 8, the choice should be the same in the two cases, problem II being derived from problem I by adding $-c$ to the entries in the second column.

This, incidentally, is also an objection against the Laplace criterion which associates the same desirability index with A'_1 and A'_2; for every 'cautious' decision-maker, however, the two actions are not equal desirable. We shall return to this point later on.

The usual justification for axiom 8 is that it follows from a presumably plausible postulate. To quote from Luce and Raiffa: 'Consider a probability mixture of two d.p.u.u.'s with the same sets of actions and states. If the second d.p.u.u. has payoffs which do not depend upon the act chosen, then the optimal set of the mixture problem should be the same as the optimal set of the first d.p.u.u.'[9]

Let the two problems be:

I

	S_1	S_2
A_1	1	4
A_1	2	2

II

	S_1	S_2
A_1'	a	b
A_2'	a	a

and suppose that after P has made his choice between A_1 and A_2, a chance mechanism determines with probabilities respectively p and $1 - p$ which of the two matrices will be used. Now, if P adopts the view that his choice ought to depend entirely on the problem I and if the procedure is to be repeated several times, the matrix of the mixed problem can be calculated in terms of expected utilities;

	S_1	S_2
A_1	$1 \cdot p + a(1 - p)$	$4 \cdot p + b(1 - p)$
A_1	$2 \cdot p + a(1 - p)$	$2 \cdot p + (1 - p)$

which shows that it is permissible to add a constant to the entries in each column of the first matrix (multiplying the matrix by a constant $p > 0$ is, of course, allowed). But are we still allowed to deduce in this way the axiom 8 from the requirement quoted above, if the choice of is to be made only once? Does the calculation of expected utilities of type '$2p + a(1 - p)$' make sense in such a case? If not, then the construction of 'expected utility matrix' for the mixed problem is not possible and axiom 8 would not follow. This, of course, applies to the 'non-repetitive' situations of choice.

As regards the axiom 11, the reason for adopting it are not clear enough. To borrow an example from Luce and Raiffa's book: consider two problems:

I

	S_1	S_2	S_3	S_4
A_1	6	2	2	2
A_2	0	5	5	5

II

	S_1	S^*
A_1	6	2
A_2	0	5

According to the axiom 11, these two problems are to be treated as equivalent, for otherwise S^* would be considered more likely than S_1 – which contradicts the principle of complete ignorance. Now, it is not clear why it should be so. On the contrary, it might be said that 'lumping together' of identical columns does not seem compatible with an implicit assumption underlying the d.p.u.u., i.e. – that no probability distribution over the set S is *a priori* excluded. In the example under discussion, the deletion of repetitious column, say s_3 and S_4, seems equivalent to assuming that the probability distribution over the set $\{S_1, S_2, S_3, S_4\}$ is (if it exists) of the type $\{p, 1 - p, 0, 0\}$. This certainly is a restriction upon the set of all possible distributions over the original set of the states of Nature.

If we are not prepared to accept axioms 8 and 11, then the two criteria left for consideration are maximin and Laplacean, for they both satisfy axioms 1–7, 9, 10. (And it should be noticed that the adoption of axiom 8 would make the Laplace criterion the only one acceptable. As has been proved by Chernoff the Laplace criterion is defined in the strict sense of the word by the axioms 1–10.)

A few words about a well-known objection against the use of Laplace criterion are perhaps not out of place here. This criterion can be interpreted as recommending the maximizing of expected utility under the assumption of equidistribution over S. Now, it is argued that different results are arrived at for different enumerations of all possible and mutually exclusive states of Nature. Thus suppose that in some case such list consists of two states: S and S'; then, under the assumption of equidistribution, the probability of S' is 1/2. But if S is split into two states S_1 and S_2, then the list consists of S_1, S_2, S' and the probability of S' is 1/3. (It remains an open question how the utility function is redefined for the new problem.) There seems to be no general rule for solving such contradictions, although in a number of fairly simple cases (e.g. games of chance) it is quite obvious how the enumeration of states of Nature 'should' be done.

It should be noticed, however, that if we want to explain the expression 'rational choice of decision' in narrower sense (to be explained below) the above objection does not apply. It seems, namely, that there are two interpretations of the problem of choice of action. According to the broader one, the only data of the problem is the set α of possible actions. Then the problem consists in (1) enumerating all the relevant and mutually exclusive states of Nature, (2) defining utility function for every pair: action – state of Nature, (3) choosing an action from the set α by applying a criterion of choice to the utility matrix constructed in (1) and (2). Let us call both (1) and (2) 'description of the situation of choice'; it may be thought that some such descriptions are adequate (and some are not) with respect to the real situation the decision-maker finds himself in. Without clarification the meaning of the word 'adequate' in this context, we may sometimes encounter difficulties in applying the theory of decision making to 'real' situation of choice.

In the narrower sense, however, the problem of rational choice: given a description of the situation of choice the form of a utility matrix (whether adequate or not, it does not matter), how (i.e., according to what criterion) the choice of action is to be made? Here the meaning of the expression 'rational choice' is restricted to the relation between the given matrix and the action chosen; it does not cover the process of constructing this matrix. If we interpreted the problem in this narrow way, the question of the proper listing of the relevant states of Nature does not arise.

Let us come back to the problem of appraising the Laplacean and maximin criteria. If they both pass the test of the set of axioms (1–7, 9, 10) adopted, it seems reasonable to look at them from another point of view.

Now, it is a well-known fact that the maximin criterion has this paradoxical property that in some cases it does not allow of improving the choice in the light of experimental evidence concerning the likelihood of the states of Nature. Thus, for instance, in the d.p.u.u.

	S_1	S_2
A_1	4	1
A_2	3	2

A_2 is the action chosen by maximin. Suppose now that an experiment is performed which can result in one of the two outcomes: x_1 or x_2; and suppose that $p(x_2/S_2) \approx 1$, therefore $p(x_1/S_2) = 1 - p(x_2/S_2) \approx 0$. If the outcome x_1 actually occurs, it would hardly seem reasonable to choose still A_2 in spite of the fact that the experiment has almost excluded S_2. On the contrary, it would seem reasonable to make the choice under the assumption (made highly plausible by experimental evidence) that the true state of Nature is S_1; and therefore choose A_1. According to maximin, however, A_2 is always better than A_1, whatever the evidence of the kind described: only if S_2 were positively ruled out (which can never happen in the case of tests of probabilistic character), would the choice be changed.

It is easy to see that this is always the case when the utility matrix has a saddle-point, that is such an entry u_{rs} which is the minimum element in its row and the maximum element in its column: $u_{rs} = \min_j u_{rj} = \max_i u_{is}$. No decision function $d(x) = A_i$ (see p. 115) is better in maximin sense than the action A_r. This is not difficult to prove. Bearing in mind that every decision function induces a probability mixture, say ξ_j of the entries in j-th column ($j = 1, \ldots, n$), it is enough to show that $u_{rs} \geq \sum_i [u_{is}\,\xi_s(u_{is})]$ for every possible distribution ξ_s. Which is obvious:

$$u_{rs} = u_{rs} \cdot \sum_i \xi_s(u_{is}) = \sum_i [u_{rs} \cdot \xi_s(u_{is})] \geq \sum_i [s_{is} \cdot \xi_s(u_{is})]$$

It thus appears that if the utility matrix has a saddle-point, then no experimental evidence, however compelling, can influence the choice made according to maximin criterion: the decision-maker might as well dispense with making experiments that could bring some information about the chances of the states S_j to occur. This seems so paradoxical that one might wish to exclude such a way of making decisions, by adopting a special postulate:

A1. For every d.p.u.u. with finite payoff, there exists such a set of probability distributions ξ_j over the entries in j-th column that a new action yielding the payoff $\sum_i [u_{ij} \cdot \xi_s(u_{ij})]$, $j = 1, 2, \ldots n$, belongs to the optimal set $\hat{\alpha}$.[10]

This requirement, violated by the maximin criterion if the case of $\xi(u_{rs}) = 1$ is ruled out, is rather weak in this sense: it only excludes the theoretical impossibility of improving the choice by experimentation; it certainly does not say that for every kind of experiment and for every set of probability distributions $p(x \mid S_j)$ over its results the information provided will enable the decision-maker to improve his choice. It is an open question whether, if under some specific assumptions (e.g. that the states of Nature are interpreted as possible values of parameter p in binominal distribution and the experiment consists in taking a sample of n independent elements) there exists a decision function belonging to $\hat{\alpha}$.

It is obviously that the Laplacean criterion satisfies A1 and if one wants to adhere to the postulates 1–7, 9, 10, A1, then this is the only criterion left for consideration. It might be recalled here that some people object against maximin on the ground that it is 'unduly pessimistic'. To make it more clear, let us consider the simple problem:

	S_1	S_2
A_1	0	1
A_2	x	x

$0 < x < 1.$

A_2 is chosen according to maximin, for all values of x. This is equivalent to maximizing the expected utility under the assumption that $p(S_1) > 1 - x$. Now, if x is small, e.g. 0.01, it follows that the 'maximiner' ascribes (implicity) to the worse state of Nature the probability higher than 0.99. The 'pessimistic character' of maximin consists precisely in this. Therefore, it is sometimes argued, maximin is not 'elastic' enough in not taking into account other characteristics of action A_i than its minimum utility value. What is a sensible procedure in games against an intelligent opponent may not be sensible in making decision under uncertainty.

We now come to the problem of appraisal of the Laplacean criterion. The fact that it satisfies a certain number of very plausible requirements does not exclude the possibility of its possessing some undesirable properties. We have already hinted (in connection with axiom 8) at a possible objection to this criterion. Let us take as a starting-point the problem

	S_1	S_2
A_1	0	0
A_2	$-c$	c

$c > 0.$

Obviously, there are three attitudes possible for decision-maker faced with such problem: I to prefer A_1 to A_2 (maximin, Hurwicz for $\alpha > 1/2$); II to prefer A_2 to A_1 (Hurwicz for $\alpha < 1/2$); III to be indifferent between A_1 and A_2 (Laplace, Savage, Hurwicz for $\alpha = 1/2$). The attitude I can be called 'cautious' ('prudent' would be the word if Prof. Braithwaite had not already given it somewhat different meaning[11]), and if someone wants to be cautious in this, as yet undefined sense, then he is bound to consider it a weakness of the Laplacean criterion that it does not discern between action A_1 and A_2. Two problems arise: 1. How to rule out the criteria violating 'cautiousness' (e.g. Laplacean or minimax regret criteria) for the general case of utility matrix $\|u_{ij}\|$. 2. Which criterion will satisfy such a requirement; it obviously could not be the Laplacean criterion.

The postulate for cautiousness might be formulated in different ways, but it essentially amounts to this: the choice-set $\hat{\alpha}$ ought not to be, in general, invariant under such transformations of the utility matrix which change the security levels of actions. By 'security level of action A_i' we mean (borrowing this expression from Luce and Raiffa) the maximin index: $\min_i u_{ij}$. It seems obvious that the decision-maker, if he is cautious at all, ought not to be indifferent to changes in security levels of admissible actions. It would not be true to say that if he adopts this attitude he is necessarily a: maximiner'. This means only that he makes his choice depend to some extent upon the security levels of the actions in α.

Now, there is more than one transformation that does not (in general) leave the security level unchanged, while preserving some other properties of the matrix. But perhaps the most simple one is that of adding a constant to all the entries in a column. We are thus lead to adopt the negation of axiom 8. This may seem a rather bold step to take; we have already seen, however, that if we want to preserve the rationale of the maximin criterion, axiom 8 must be rejected. True, that it is one thing not to accept an axiom, and quite another – to adopt its negation. But by negating axiom 8 we are not implying that adding a constant to a column always influences the choice of action. On the contrary, the suggested postulate is rather weak:

A2. For some d.p.u.u.(from which dominated actions have been eliminated) there is such a column j and such constant c that adding c to each entry in the column j does change the choice-set $\hat{\alpha}$.

This postulate, although very weak, rules out the Laplacean criterion. It implies that in some cases lowering the security-level of an action, previously optimal, makes it non-optimal (and vice versa). It seems that the postulate for cautiousness could hardly be expressed in a weak form.

Now, if we want to adopt postulates 1–7, 9, 10, A1, A2, a possible solution is to mix the maximin and Laplacean index with some parameter β. Formally, the criterion would recommended choosing the action which maximizes

$$m_i \beta + L_i(1 - \beta),$$

where

$$m_i = \min_j u_{ij}, \; L_i = \frac{1}{n} \sum_j u_{ij}, \; 0 < \beta < 1.$$

It is easy to see that this criterion satisfies the above mentioned requirements. We shall prove, for instance, that it satisfies axiom 9 (convexity of $\hat{\alpha}$). Let us consider two actions:

	S_1	S_2		S_n
A_1	x_1	x_2	....	x_n
A_2	y_1	y_2	....	y_n

and suppose that they both belong to $\hat{\alpha}$, i.e. their β-indexes are equal:

$$m_1\beta + L_1(1-\beta) = m_2\beta + L_2(1-\beta). \tag{*}$$

The two actions are to be mixed by $(\eta,\ 1-\eta)$. Denote by m and L the maximin and Laplacean indexes of the mixture:

$$m = \min_i[x_i\eta + y_i(1-\eta)],\ L = \frac{1}{n}\sum_i [x_i\eta + y_i(1-\eta)]$$
$$= L_1\eta + L_2(1-\eta).$$

In view of $0 \neq \beta \neq 1$, there are two possible ways for the equation (*) to be fulfilled:

1. $m_1 = m_2\,;\ L_1 = L_2\,.$

   Then $m = \min_i[x_i\eta + y_i(1-\eta)] \geq m_1 = m_2\,,$

   and $L = L_1\eta + L_2(1-\eta) = L_1 = L_2;$

   therefore $m\beta + L(1-\beta) \geq m_1\beta + L_1(1-\beta) = m_2\beta = L_2(1-\beta).$

2. $m_1 \neq m_2\,;\ L_1 \neq L_2\,.$

   Suppose, without loss of generality, that $m_1 > m_2$ and $L_1 < L_2$. From (*):

   $$\frac{m_1 - m_2}{L_2 - L_1} = \frac{1-\beta}{\beta};$$

   therefore $m_1 - m_2 = c(1-\beta)$; $L_2 - L_1 = c \cdot \beta$, where $c > 0$.
   The smallest possible minimum in the mixed action is:

   $$m' = m_1\eta + m_2(1-\eta) = m_2 + c(1-\beta)\eta,$$

   and the Laplacean index of the mixed action is:

   $$L = L_1\eta + L_2(1-\eta) = L_1 + c\beta(1-\eta).$$

   Thus the smallest possible β-index for the mixed action is:

   $$m'\beta + L(1-\beta) = m_2\beta + L_1(1-\beta) + c(1-\beta)\beta.$$

   But in view of $m_1 = m_2 + c(1-\beta)$ and $\beta \neq 0$ we have:

   $$m_1\beta + L_1(1-\beta) = m_2\beta + c(1-\beta)\beta + L_1(1-\beta).$$

Thus the smallest possible β-index of the mixed action is equal to the β-index or original actions, which shows that mixing of optimal actions does not lower the β-index. Therefore the set $\hat{\alpha}$ is convex.

The choice of the actual value of β is left open to the decision-maker P. It expresses his 'degree of cautiousness' (or possibly 'pessimism') in a similar way as Hurwicz's parameter α does. An empirical determination of β from P's attitude towards a simple problem would, therefore, be: 'choose such value of x that you are indifferent between A_1 and A_2 in the following d.p.u.u.

	S_1	S_2
A_1	0	1
A_2	x	x

Then β is calculated as $1 - 2x$. It should be noticed that the choice of x is restricted to $x < \frac{1}{2}$; for $x \geq \frac{1}{2} A_2$ is the action preferred, irrespective of the value of β.

This 'subjective' interpretation of β seems to be the only one possible. Every attempt to make β a function (invariant, of course, under linear transformations of utility) of some characteristic values of the matrix $\|u_{ij}\|$, breaks down because axiom 7 is then not satisfied. Some obviously properties of the β-criterion might be mentioned here.

The β-index for each action is, like the Laplacean index, a strictly increasing function of all the utility-entries in the respective row.

The choice-set $\hat{\alpha}$ is not altered by permutation of entries within each row. (Again like in Laplacean procedure but unlike in Hurwicz's and maximin procedures, some actions might be excluded from $\hat{\alpha}$ because in the transformed matrix they would be dominated).

If the choice is to be repeated several times in a given d.p.u.u. and probability mixing of the actions in α is allowed, then some randomized actions might be better (according to β-criterion) than non-randomized ones. Laplacean choice-set is, however, invariant under the extension of α to include randomized actions; thus Laplacean choice is the same for repetitive as for non-repetitive situations, which does not seems to be a desirable property. Some people, however, reject randomization altogether.

Against the β-criterion it might be argued that it does not satisfy some stronger postulates for 'cautious' choice. It does not distinguish, for instance, between A_1 and A_2 in

	S_1	S_2	S_3
A_1	-1	0	2
A_2	-1	1	1

while action A_2 seems 'safer' from the point of view adopted. In other words: as long as the security level of an action is kept constant, it does not matter how the sum total of the utilities for this action is distributed; but some people may prefer, of two actions having the same total sum of utilities, the one with more 'uniform' distribution. It seems difficult, however, to devise a criterion which would conform to this preference; introducing a 'variance' of the utility function would lead to contradiction in view of the axioms of utility adopted (cf. Luce and Raiffa, op. cit. p. 32; fallacy 2).

To sum up, the notion of 'rational choice' in d.p.u.u. is imprecise enough so it allows different explications. It is true that some postulates for rational choice have been formulated, which makes it possible to appraise the suggested explications in their several aspects. All such postulates, however, cannot be reconciled and it remains a matter of individual preference which of them are to be taken as basic. I have chosen a set of postulates (not without some justification, I hope) and suggested a criterion satisfying them. It is obvious that from some other point of view the set of axioms and the criterion adopted might not be found satisfactory.

NOTES

[1] 'Utility' is thus a function defined on a set of what we call here 'situations'. Such a function must satisfy certain conditions. There exist several systems of axioms for the utility-functions: the axiomatic treatment of utility we are assuming here is that given by Luce and Raiffa in *Games and Decisions: Introduction and Critical Survey*, J. Wiley, N. York, 1957. The empirical interpretation of the concept can be found in the book.

One of the most important consequences of the axioms we are adopting is that the utility function is defined only up to a (positive) linear transformation, i.e. if u is the utility function for P, so is $a \cdot u + b$, where $a > 0$ and b is any number. It follows, for instance, that all the problems of choice defined by a matrix

	S_1	S_2
A_1	0	0
A_2	$-c$	c

are equivalent, whatever the value of c may be.

[2] For an exposition of the purposes and principal results of praxeology see: T. Kotarbiński, *Traktat o dobrej robocie* [*A Treatise on Good Work*], Łódź 1955.

[3] Cf. for instance paper by H.A. Simon, *A Behavioral Model of Rational Choice* reprinted in his book *Models of Man*, J. Wiley, New York 1957.

[4] D. Blackwell and M.A. Girshick: *Theory of Games and Statistical Decisions*, J. Wiley, New York 1954.

[5] Cf. L.A. Goodman, *On Methods of Amalgamation in Decision Processes* ed. by R.M. Thrall, H. Coombs and R.L. Davies, J. Wiley, New York 1954.

[6] I am leaving out the justification of the axioms, which is given in full in the book I am quoting from.

[7] If axiom 5 is adopted, a simplification of axiom 4 is possible. Axiom 5 excludes the possibility that $A \in \hat{\alpha}$ and $A' > A$ (for some A'). It is thus enough for axiom 4 to assert that: If $A \in \hat{\alpha}$ and $A \sim A'$, then $A' \in \hat{\alpha}$. A rather obvious requirement.

[8] It is assumed that maximin and Hurwicz criteria are applied to d.p.u.u. after all the inadmissible have been eliminated from α. Otherwise axiom 5 would not be satisfied by these criteria.

[9] Op. cit., p. 290.

[10] The trivial case is, of course, excluded of such a set $\xi_1, \ldots \xi_n$ which would make the new action identical in payoffs with one of the old actions. Incidentally, the Hurwicz's α-criteria do not satisfy $A1$; a suitable matrix is easy to construct.

[11] Cf. R.B. Braithwaite, *Theory of Games as a Tool for the Moral Philosopher*, Cambridge 1955.

15

THE CONCEPT OF DISTRIBUTION OF GOODS

In this paper I wish to consider the following problem: what is required of the distribution of n objects among m persons? The term 'object' is interpreted here widely as possible. It is assumed that we know how the objects subject to distribution are appraised by different persons. It is also assumed that they are desired (in the sense defined below), and hence may be called goods.

The problem formulated in this way is related to the issue of distributive justice. The relationship is not very close, since it can hardly be expected that the meaning of the concept of justice could be explained by reference to such a simple model. The data are too scanty. Many essential circumstances which affect nearly every distribution of goods, and on which an ethical appraisal of such a distribution depends, are disregarded here. Moreover, nothing is assumed about the participants in the distribution except for the knowledge of their preferences; nor is the social context, which might provide some recommendations for the intended distribution, defined in any precise way.

The shortcomings of such simplifications are obvious. Their advantage consists in the possibility of a fairly precise formulation of the basic underlying intuitions without becoming involved in the difficulties due to the variety of real conditions. Such idealizations are useful, as has been shown by research done over the last years.[1] Normative models of human behaviour have made it possible, for instance, to examine the consistency of certain elementary praxiological and ethical requirements.[2] Simplification is the price which it is worth paying for such results.

A PRELIMINARY DESCRIPTION OF THE MODEL

It is assumed that none of the objects subject to distribution is divisible. This assumption differs from that usually adopted in the literature of the subject. Thus, for instance the 'fair division games', to be discussed at the end of the paper, refers to the situations in which the good being distributed is continuous in nature and is infinitely divisible. It seems that distributions of sets of objects occur sufficiently often in practice. This to my knowledge, has not been done so far.

The valuation of objects by the various individuals is interpreted in the weakest possible way, namely as an ordinary preferential ordering (which does not exclude indifference). The usual assumption made in the literature of the subject – for instance, in the theory of games – is that of the knowledge of intensity of preferences, which makes it possible to speak about the ratios of differences in valuations (the interval scale). This is obviously a stronger assumption, because location of objects on an interval scale implies their preferential ordering, but not *vice versa*. The adoption of a weaker assumption may be justified in two ways. First, it is more realistic: empirical confirmation of intensity of preferences is extremely difficult even under laboratory conditions,[3] not to speak of much more complex conditions occurring in real-life situations. Secondly, making use of a stronger assumption could consist only in an operation that is hardly admissible, namely an interpersonal comparison of valuations. If that operation is forsaken (which has important arguments in its favour), then the knowledge of intensities of preferences becomes superfluous in our case.

The final remark concerns the probabilistic interpretation of the distribution. It seems indispensable to include probability distribution in the definition of (generalized) division; as is commonly known, it is only through randomization that we can ensure the satisfaction of the intuitive requirements of egalitarianism, that is, simply of symmetry. This is of particular importance when the number of (indivisible) goods is smaller than that of participants in the distribution. But it is not only in this case that randomization is the only means of 'equalizing' distribution.

These remarks justify, I hope, the introduction of the following concepts.

FORMULATION OF THE PROBLEM

There are m persons called participants in a distribution ($m \geq 2$), who are symbolized by $S_1 \ldots, S_m$.

There are n objects ($n \geq 1$), which are symbolized by $A_1 \ldots, a_n$ or $A, B, C, \ldots$ For formal reasons it is convenient to add to them a special $(n+1)$-th object, to be symbolized by 0. If a person is assigned to the object 0, we interpret it that he is assigned nothing. The set of all objects subject to distribution is symbolized $\mathfrak{R} = \{0, A_1, \ldots, A_n\}$

It is assumed that it is known how a person S_i values the objects that are elements of the set $\mathfrak{R}$. To put it more strictly, for every i ($1 \leq i \leq m$) there is given a preference ordering of the objects that are elements of the set $\mathfrak{R}$, from the most to the least desired one. That order is reflected by the relation $\leq$. The formula $A \leq_i B$ is read: the object B is for the person S_i at least as much desired as is the object A. Thus every relation $\leq_i$ is a weak ordering defined on the set $\mathfrak{R}$.

It is convenient to introduce the relation of strong preference:

$$A <_i B =_{df} A \leq_i B \wedge \sim(B \leq_i A).$$

Every set of preferences orderings $\{\leq_1, \ldots, \leq_m\}$ is called *profile of preference orderings.*

Let us assume that the profiles of preference orderings satisfy the following condition: For every i and every j: $0 <_i A_j$ $(1 \leq i \leq m,\ 1 \leq j \leq n)$. In other words, every person prefers an arbitrary object A_j to the *quasi*-object 0. In that sense the object $A_1, \ldots, A_n$ may be called goods. Notice that this assumption permits to disregard 0 when different profiles of preferential orderings are considered.

Division means associating every A_i with exactly one person, on the condition that the person with whom no A_i is associated receives the object 0. Hence, if the object 0 is treated as the empty set, division is an m-element *quasi*-partition of the set $\{A_1, \ldots A_n\}$. The term 'quasi-partition' is used because the nonemptiness of the elements of that 'partition' is not postulated.

Distribution means any distribution of probability on the set of all divisions. Distribution so defined is, of course, a generalized concept of division, since every division may be treated as a distribution, namely such in which the probability of that division equals 1.

Distribution rule associates a definite distribution with every profile of preference orderings. Thus, if R is the set of all possible orderings of the set $A_1, \ldots, A_n$, then the distribution rule is a function defined on the Cartesian product

$$\underbrace{R \times R \times \ldots \times R}_{m \text{ times}},$$

which takes on values from the set of all possible distribution.

This concept can be illustrated by a simple example. Suppose that $m = n = 2$, so that we have to do a division of two objects, A and B, among two persons S_1 and S_2. The valuations of the two objects can be done in three ways: A may be valued more than B, or conversely, or both may be valued equally. Thus there are $3 \cdot 3 = 9$ possible profiles of preference orderings. Further, the set of all divisions has the following four elements: 1) S_1 receives A while S_2 receives B; 2) conversely; 3) S_1 receives A and B, while S_2 receives 0; 4) conversely.

The following table shows one of the (infinitely many) distribution rules possible in such a case. The rows correspond to the profiles of preference ordering, defined in the first two columns. The numbers occurring in the next four columns stand for probabilities, the sum of which in every row must be 1, since for every profile of preference orderings those numbers constitute the probability distribution on the set of the four possible divisions.

	S_1	S_2	A/B	B/A	$AB/0$	$0/AB$	
1	$A > B$	$A > B$	0.5	0.5	0	0	
2	$A > B$	$A = B$	1	0	0	0	
3	$A > B$	$A < B$	1	0	0	0	
4	$A = B$	$A > B$	0	1	0	0	

5	$A = B$	$A = B$	α	$1-\alpha$	0	0	$(0 \leq \alpha \leq 1)$
6	$A = B$	$A < B$	1	0	0	0	
7	$A < B$	$A > B$	0	1	0	0	
8	$A < B$	$A = B$	0	1	0	0	
9	$A < B$	$A < B$	0.5	0.5	0	0	

Obviously, the case of $m = n = 2$ is trivial. Note, however, that a complete presentation of a more complex case would be inconvenient and not easy to read: for $m = 3$, $n = 2$ the table would have to have 27 rows and 9 columns, and for $m = 2$, $n = 3$, 169 rows and 8 columns.

The distribution rule used as the example, and defined by the table above, has been constructed so as not to disagree with intuitive requirements associated with a 'fair' division. The point, however, is how such intuitive requirements can be made more precise.

It may be stated provisionally that a distribution rule is supposed not to work in favor of any person (egalitarianism) and to assign to the participants in the division those goods which they value most, to the extent permitted by the situation (optimality). We shall analyze these two requirements.

EGALITARIANISM

In its strongest version (which consequently cannot usually be put into practice) the requirement of egalitarianism is that the objects assigned to various persons should be equally valued by them. A distribution rule complying with that requirement would have to assign zero probability to all those distributions which assign to various persons objects that those persons do not value in the same degree. This is obviously impossible when $m > n$, and also in all those cases when the preference of the participants in the division are identical or at least sufficiently alike; the resulting conflict of interests excludes 'perfect equality'.

We can of course, easily imagine such a profile of preference orderings in which the above requirement can be satisfied. Take, for instance, for $m = n = 3$:

S_1	S_2	S_3
A	B	C
B	C	A
C	A	B

(Objects A, B, C are here listed in the order of decreasing preference: e.g. S_1 values A most, then B, and then C). Note that the requirement of 'perfect equality' could be satisfied by assigning to persons S_1, S_2, S_3 the objects A, B, C respectively, or the objects B, C, A respectively, or the objects C, A, B, respectively. Obviously, the first of the these three divisions would be optimal.

But if the requirement of egalitarianism is expected to work in practice in a *general* way, that is, for all profiles of preferential orderings, then the existence of conflicting preferences forces us to reduce our requirements concerning the distribution rule. The concept of probability which makes it possible to refer to 'equal opportunities' is helpful here.

Before, however, referring to that concept, let us notice that the requirement of absolute equality can be saved to some extent. We can exclude at least those divisions (by assigning them zero probability) which are nonsymmetrical 'over and above necessity'. For instance, for $m = n = 2$, the division that assigns objects A and B to the person S_1, and the object 0 to the person S_2. More generally, we exclude those divisions in which a person S_i receives k objects, and a person S_j $k + 2$ objects. It is self-evident that such a requirement can always be complied with. It is satisfied, for instance, by the distribution rule formulated on p. 130. Let us now revert to the formulation of the probabilistic version of egalitarianism. In order to avoid complicated statements that could not contribute anything new to the conceptual aspect of the problem, we shall confine ourselves to the case when $m = n$. It is also assumed that all preference orderings are strong, so that there are no objects that are valued in the same degree. Generality of the analysis is thereby not impaired, for it can easily can be seen that if the requirement of egalitarianism is satisfied for strong preference orderings, it is *a fortiori* satisfied for weak orderings, which leave more freedom in the choice of the distribution rule. Here is, in my opinion, the natural formulation of the requirement of egalitarianism, conceived as 'equal opportunities'. For every person we number the objects subject to division in the order of decreasing preferences: the object most valued by a given person is marked with number one, the next number with number two, *etc.* Now the requirement will be that *every participant in the division should have the same probability of receiving the k-th object* ($k = 1, 2, \ldots, n$).

This is illustrated by a simple example. Let the profile of preference orderings, for $m = n = 3$, be as follows:

S_1	S_2	S_3
A	B	B
B	A	C
C	C	A

The class of permissible distributions is shown by the following table.

1	2	3	4	5	6
ABC	ACB	BAC	BCA	CAB	CBA
x_1	x_2	x_3	x_4	x_5	x_6

Divisions are marked here according to the numbering of persons; thus, division (4) assigns the object B to the person S_i, C to S_2, and A to S_3. The numbers x_i stand for probabilities, hence $x_i \geq 0, \sum x_i = 1$.

The requirement of egalitarianism, interpreted as equal opportunities, imposes certain conditions upon those probabilities, namely:

$$\begin{aligned} x_1 + x_2 &= x_1 + x_6 = x_2 + x_5 \\ x_3 + x_4 &= x_3 + x_5 = x_1 + x_3 \\ x_5 + x_6 &= x_2 + x_4 = x_4 + x_6 \end{aligned}$$

For instance, the questions in the second line originate from the fact that $x_3 + x_4$ is the probability of the person S_1 receiving the object B, which for him stands second in the order of preference. Likewise, $x_3 + x_5 =$ the probability of the person S_2 receiving the object A, which also is for him second in the order of preference. Finally $x_1 + x_3 =$ the probability of the person S_3 receiving the object C, which for him is the second in the order of preference. By the requirement formulated above, these probabilities are equal. The above system of equations can easily be reduced to the following ones:

$$\begin{aligned} x_1 &= x_4 = x_5 \\ x_2 &= x_6 \end{aligned}$$

The meaning of these equations becomes obvious when in notiations of various divisions the names of the objects are replaced by their numbers in the order of preferences of the respective persons:

x_1	112
x_4	233
x_5	321
x_2	131
x_6	313
x_3	222

It can be seen that equal probabilities separated by lines from one another guarantee equal opportunities.

The problem arises whether the requirement of equal opportunities can be satisfied for every profile orderings. The answer is yes. It suffices to notice that the condition which in the light of that requirement is satisfied by the distribution associated with a given profile of preferences takes on the form of a system of equations analogous to that given above. The consistency of such a system is guaranteed by the fact that it is satisfied for $x_i = 1/n!$. Hence the requirement of equal opportunities is acceptable.

This requirement does not, as a rule, univocally determine the distribution rule, but it restricts its choice. Notice that in the special case of the maximally conflicting preferences (i.e., preferences identical for all participants in the division) the requirement of equal opportunities determines the distribution rule in an univocal manner, namely $x_i = 1/n!$. (The proof, since it is very simple, is omitted.) This fact deserves notice, because in the analyses concerned with distributive justice it is usually assumed that the order of values is the same for all the participants in the division. Now, under such an assumption only the ideal symmetry of the distribution of probabilities satisfies the requirement of egalitarianism in the version adopted in the present paper.

OPTIMALITY

The intuitions associated with this concept are much more difficult to formulate precisely than the requirements of egalitarianism. The point is to give participant in the division a maximum share, within the limits determined by a given situation, hence also by the profile of preference orderings. But the word 'maximum', used above, will have a definite meaning only if there is a relation ordering the set of all divisions from the point of view under consideration. Such a relation could be described as the division P_1 is for the participants at least as advantageous as is the division P_2. Such a relation would have to be connected and transitive, since it is to order the set of all divisions.

But here a difficulty arises, because we do not know whether coherence of such a relation could be ensured without resorting to obviously arbitrary decisions. It is easy to indicate two divisions that seem incomparable in terms of the relation referred to above.

To revert to the last example. Which of the two divisions, *ACB* or *BCA*, is more advantageous to the participants with such a profile of preference orderings? The former is characterized by the numbers 131, the latter by 222. We may compare those numbers by summing them: $1 + 3 + 1 < 2 + 2 + 2$. That would indicate that the first of the two divisions is more advantageous than the second. But such an operation would assume that assigning to the person S_i an object less valued in his preference scale can be compensated by assigning to the person S_j an object more valued in his preference scale. Such an operation, called interpersonal comparison of utilities, is exposed to the same objections as Bentham's calculus of pleasure and pain.[4] Such arithmetic does not seem admissible. On the other hand, however, there is no other way of ensuring the coherence of the relation: "the division P_1 is more advantageous to the participants than the division P_2."

Accordingly we have to renounce the coherence and to interpret the above relation so that it may at least be a *partial* ordering in the set of the divisions. This can be done, and as a result we obtain the weakest possible version of the requirement of optimality. The starting point is that there are pairs of divisions such that one division is indisputably better than the other. For instance, the division characterized by the number 112 is certainly more advantageous to the participants than that characterized by the number 222, since in the former case every participant receives an object at least as valuable for him as in the letter case, and some participants (S_1 and S_2) obtain objects they value more.

Borrowing terminology from theory of games we will say that the division characterized by the numbers $k_1, k_2, \ldots, k_n$ is *dominated* by the division characterized by the numbers $l_1, l_2, \ldots, l_n$ if for every $i: k_i \geq l_i$. A division which is not dominated by any other division will be called admissible.

Now the *weakest* version of the requirement of optimality (and probably the only one that is not open to objections) excludes inadmissible divisions. In other words, the requirement of optimality so conceived demands from the distribution rule that *inadmissible divisions will be assigned zero probability.*[5]

EGALITARIANISM VERSUS OPTIMALITY

It is remarkable that even the weakest version of requirement of optimality is inconsistent with the requirements of equal opportunities. One counter-example suffice to prove that. We can use for that purpose the last case analyzed.

We notice that division (3) is not admissible being dominated by division (1); likewise, division (6) is not admissible being dominated by division (1); finally, division (4) is dominated by (1), and thus it is not admissible. Hence by the requirement of optimality we ought to have:

$$x_3 = x_6 = x_4 = 0.$$

But, on the other hand, the requirement of equal opportunities demands that we have:

$$x_1 = x_4 = x_5 \quad \text{and} \quad x_2 = x_6$$

This jointly yields that $x_i = 0$ for $i = 1, 2, \ldots, 6$, which is obviously impossible, since, by the definition of distribution, $\sum x_i = 1$. Thus we have arrived at a contradiction.

IS IT POSSIBLE TO GENERALIZE THE REQUIREMENT OF EQUAL OPPORTUNITIES?

The requirement of equal opportunities is exposed to the objection that it interprets too narrowly the meaning of the concept of justice as applied to the problem of division. If we require from a division to be fair, we do not necessarily mean that everyone ought to have the same opportunity of receiving of what he deems valuable. We mean only that no one be privileged, as a *concrete individual.* This is not at all equivalent to the requirement of equal opportunities. It may be so that one person justly deserves more than other persons do, because of his past achievements, needs, abilities, *etc.*

This was pointed out by the late Kazimierz Ajdukiewicz in his brilliant article on the concept of justice. He sees the essence of justice in the requirement which he calls the principle of equal measure.

"By that principle, whenever a statement according a right to the person *X.Y.*, specified by name, is true, a general principle must be true which accords that right to every person that has a property which the person $X.Y.$ has, but which neither consist in this that he is Mr. $X.Y.$ nor is any such property concerning which it could be shown *a priori*, by referring to logical operations alone, that Mr. $Y.X.$ has that property."[6]

If we agree to that, we have to conclude that it is not the distribution rule itself that is just or unjust, but the manner in which it is applied. For suppose that a distribution rule privileges a concrete person S in the sense that it accords him a greater opportunity of receiving what he values. According to Ajdukiewicz's interpretation, this fact is in itself just nor unjust because it is not known whether the principle of equal measure has been violated or not. That could be seen only if it were revealed by what criterion the person S has been privileged.

It follows that the possibility of privileging any participant in a division with respect to other participants is an indispensable condition of the applicability of Ajdukiewicz's principle of equal measure to the model of the division of goods as analyzed in this paper. In other words, the condition imposed on the distribution rule ought to have such a parametrical form that the opportunities of the participants in the division of receiving goods could have arbitrary ratios. What is that ratio, is decided by reasons independent from the analyzed model. That ratio of opportunities would be just, or unjust, according to whether it were determined in accordance with the principle of equal measure, or not.

We shall now try to generalize the requirement of equal opportunities by imparting it such a parametrical form. Let $p_i(k)$ stand for the probability of the i-th person receiving the object on the k-th place in this order of preference.

Let, for every k, $p_1(k) : p_2(k) : \ldots : p_m(k) = a_1 : a_2 : \ldots : a_m$. The (positive) numbers a_i are the weights associated with the various persons by a criterion which remains outside the formal structure of the model. In the special case when $a_1 = a_2 = \ldots = a_m$ the requirement of equal opportunities is satisfied.

It turns out, however, that freedom in assigning arbitrary weights to various participants in the division is illusory because the following theorem holds: the condition formulated above is satisfied *only if* $a_1 = a_2 = \ldots = a_m$.

The general proof of the theorem is inconvenient to supply here, and at the same time sufficiently simple not to be quoted here. Its structure can easily be shown by an example. Assume the profile of preference orderings recorded on p. 132. Suppose that the weights assigned to the persons S_1, S_2, S_3 are, respectively, a, b, c. Hence

$$p_1(k) : p_2(k) : p_3(k) = a : b : c \quad k = 1, 2, 3.$$

Complying with the symbolism adopted previously we write down this condition in the form of $3! = 6$ equations:

$$(x_1 + x_2)\, b = (x_1 + x_6)\, a \tag{1}$$

$$(x_1 + x_2)\, c = (x_2 + x_5)\, a \tag{2}$$

$$(x_3 + x_4)\, b = (x_3 + x_5)\, a \tag{3}$$

$$(x_3 + x_4)\, c = (x_1 + x_3)\, a \qquad (4)$$

$$(x_5 + x_6)\, b = (x_2 + x_4)\, a \qquad (5)$$

$$(x_5 + x_6)\, c = (x_4 + x_6)\, a \qquad (6)$$

By summing up the equations (1), (3) and (5) by sides and taking into account that $\sum x_i = 1$, we obtain $b = a$. By summing up the equations (2), (4) and (6) we obtain in the same way $c = a$. Hence $a = b = c$.

There is no difficulty in generalizing this reasoning in order to cover an arbitrary m and an arbitrary profile of preference orderings. The principle remains the same, and consist in eliminating probabilities by summing properly structured equations.

The conclusion from this theorem is as follows. We have stated above that the possibility of assigning different weights to the various individuals is an indispensable condition of applying the principle of equal measure to the model of division of goods as analyzed in the present paper. This, however, proved impossible. Consequently, the principle of equal measure is not generally applicable to such a model of division of goods. The only form in which the requirement of justice is applicable is the requirement of equal opportunities.

Why is this so? This is, I think, due to assumption of indivisibility of the goods to be distributed. Such a difficulty would not arise if the assumption were modified, for instance, as follows. A single good which is infinitely divisible is subject to division. The preferences of various persons are identical and are reflected by a strictly increasing function defined on the measures of quantities of that good. (Such a situation is approximated by the division of a given sum of money among a number of persons.) It can easily be seen that under such assumption the degree of a relative privileging of the participants in the division can be controlled by will. And that 'relative privileging' is in turn subject of appraisal from the point of view of the principle of equal measure.

RELATED STUDIES

The problem of division has already been an object of formal studies, which deserve attention. A more detailed analysis is to be found in the book by Luce and Raiffa.

First of all, there is a number of studies concerning the games of fair division. Suppose that there is homogenous and infinitely divisible good, which is to be divided among m persons. The task is to construct an m-person game such that its participants, acting in their respective interests, achieve an equal, and in this sense fair, division. For the simplest case when $m = 2$, the solution consist in the procedure commonly known as 'divide and choose', where one person divides and the other chooses. Generalization for an arbitrary m is due to B. Knaster and S. Banach. When we deal with a set of indivisible objects the suggestion is that we resort the valuations in terms of money and pay out the differences (positive and negative).

A just division is, in that case, interpreted in an egalitarian manner as assigning to all participants in the division equal parts of the good to be divided. The concept of probability is not necessary here, as the divided good is continuous. The point is to set up a social mechanism that would 'automatically' ensure egalitarianism under the assumption of strictly egoistic motivation on the part of the persons concerned.

The problem of arbitration in non-zero sum games seems to be more complex from the ethical point of view. The games of that kind distinguish themselves by an *only partial* conflict of the interests of the participants, who may gain or lose something together. (For instance, in case of trading negotiations it usually happens that *both* parties prefer certain deals to no deal at all.) Hence, instead of playing the game, the parties could refer to the arbiter, who would propose the solution. The problem is, what considerations such an arbiter should take into account.

Different principles of a just arbitration are suggested and different justifications are offered in the studies on this subject (Nash, Raiffa, Braithwaite). In addition to certain requirements of consistency, all solutions include two requirements that are intuitively obvious: that the solution formulated by the arbiter be as highly valued as possible by both parties, and that it be symmetrical, if the situations of the players are symmetrical. Thus we encounter again a version of egalitarianism, which is weaker in this case because the strategic position of the players usually is not the same, which must in turn be reflected in the solution formulated by the arbiter (since on that condition it is only to be expected that both parties will accept the arbiter's decision). The problem of arbitration is a special problem of division, with certain additional assumptions that drastically reduce the number of admissible solutions. Moreover, as in all games, the results are evaluated by parties in a stronger sense than that adopted in the present paper; not only the order but also the intensity of preferences are taken into account.

The assumption made in the present paper are very similar to those to be found in the literature an a group decision making. It is assumed (Arrow, Black, Coombs) that the profile of preference orderings is given for a fixed set of persons and a fixed set of persons and a fixed set of objects. The task, however, is different: to assign to such a profile a common order of preferences. Some of the terms used in this paper have been drawn from those studies. The difference between the problems discussed is self-evident.

SUMMARY

This paper deals with the requirements for the division of goods. The goods are treated as indivisible objects. The data concerning the issue of how they are valued by participants in the division are taken in the weakest version (ordering as to 'desirability').

The concept of division, and its probabilistic generalization, called distribution, are introduced. A distribution rule associates such a generalized division with every possible system of individual preferences. The intention was to impose on the distribution rule conditions which comply with ethical intuitions.

Requirements of equal opportunities demands that the distribution rule ought, with equal probability, assign to the participants in the division those objects which they value in the same degree. The requirement of Pareto optimality demands that a division indisputably worse than some other division as to the satisfaction derived by the participants, be excluded. It turns out that requirements are, in general inconsistent.

The analysis of the 'principle of equal measure', in which Kazimierz Ajdukiewicz saw the essence of the concept of justice, led to a tentative generalization of the requirement of equal opportunities. It turns out that such a generalization is impossible.

Translated by *Olgierd Wojtasiewicz*

NOTES

* *Studia Filozoficzne*, 2. 1966. This paper was affectionately dedicated to Professor Maria Ossowska, on the occasion of the fifty years of her research on problems of ethics.

[1] *Cf* K. Szaniawski, *Cybernetyka i modele filozoficzne*, in *Cybernetyka; argumenty za i przeciw*, Warszawa, 1965, p. 54.

[2] "The point I have been concerned to make here is that the Milnor and Arrow's impossibility theorems are as philosophically relevant to the foundations of the concepts of rationality and justice as are the paradoxes of set theory for the foundations of mathematics. These impossibility theorems demonstrate that our naive intuitions about rationality and justice cannot be counted upon to yield a coherent and consistent theory". Patrick Suppes, 'The Philosophical Relevance of Decision Theory,' *The Journal of Philosophy*, Vol. LVIII, No. 21, 1961, p. 612.

[3] Convincing examples in that respect are given in *Decision Making* by Davidson, Suppes and Siegel, Stanford University Press, 1957.

[4] *Cf.* R. D. Luce and H. Raiffa, *Games and Decisions*, New York, 1957, pp. 33, 345ff.

[5] In the theory of non-zero sum games, admissibility, interpreted as above, is also called Pareto, optimality. *Cf. ibid.*, p. 118.

[6] K. Ajdukiewicz, *O sprawiedliwości*, in *Język i poznanie*, vol. 1, Warszawa, 1960, p. 372.

16
THE VALUE OF PERFECT INFORMATION

This paper is an attempt to assess the value of information a certain type, in terms of its usefulness for decision making.

It has been said often enough that information, whatever this overworked word means, has a pragmatic aspect to it, in the sense that it provides a basis for better solutions of decision problems. To quote one such opinion, ". . . the 'prediction' value of information rests upon the power it gives to the recipient to select his future action, out of a whole range of uncertainty, as to which action to take."[1]

Some writers even postulate a special concept of 'pragmatic information', in addition to the two already established uses of the word (the so-called statistical and semantic information). I do not believe that anything can be gained by ascribing a new meaning, call it pragmatic, to the word 'information'. On the other hand, it is possible to speak of the pragmatic value of certain statements or, rather, classes of statements relevant to a decision problem. Such statements will be here called 'information'. It is convenient to adopt this usage and I hope that no misunderstanding will arise, so long as it is clearly stated what we have in mind.

I shall begin by describing briefly the decision problem. Statements relevant to its solution will be called 'information, relative to the problem'. Some types of them will be singled out; the simplest type will be called 'perfect information'. Then I shall suggest some ways of assessing the pragmatic value of such perfect information.

I TYPES OF INFORMATION WITH RESPECT TO A DECISION PROBLEM

Decision problem, in its basic form, is defined as the well-known ordered triple $\langle A, S, u \rangle$ where A is the set of actions available to the decision maker, S is the set of states of Nature that determine the consequences of any action, u is a real function defined on the Cartesian product $A \times S$, representing the valuations, usually called 'utilities', of the consequences. Here, u is assumed to have the properties postulated by von Neumann and Morgenstern. It follows that u is unique only according to a linear transformation and that the expected value of utility is utility.

For the sake of simplicity, we shall restrict our attention to finite decision problems: the sets A and S will have m and n elements, respectively. Hence, the decision problem will be represented by a matrix $\|u_{ij}\|$; $i = 1, \ldots, m$; $j = 1, \ldots, n$.

In all known theories, information amounts essentially to the reduction of the set of all possibilities to its (proper) subset. If we adopt this view, different types of information concerning S can be defined as follows.

First, information concerning S may be either categorial or probabilistic. Categorial information, in turn, may or may not be precise. If precise, it specifies the actual state of Nature, and is thus of the form: $x = s_j$, which means: the actual state of Nature is s_j. If imprecise, it specifies a (proper) subset of S to which the actual state of Nature belongs: $x \in S'$, where $S' \subset S$, $S' \neq S$. Imprecise information can be classified according to its degree of precision, i.e. the number of elements in S'. Precise information is thus seen to be a limiting case of $x \in S'$, with S' containing just one element.

Probabilistic information says something about the probability distribution on S. Let us notice an important difference. In the former case x was simply a variable defined on S; it has been assumed that exactly one of the statements $x = s_j$ was true, but the decision-maker did not known which one. When we speak of probabilistic information we implicity assume the actual state of Nature to vary randomly. Therefore, X is now a random variable, and we are stressing this fact in the usual way, i.e. by using a capital letter. The set of all possibilities is now the class, say P, of all probability distributions on S. Exactly one of the distributions actually applies, but the decision-maker does not know which one. A precise information is now of the form: $y = p$, which means: the actual distribution on S is p, where $p \in P$. An imprecise information states that $y \in P'$, where $P' \subset P$, $P' \neq P$. Only some special cases of such information seem to be of interest.

The above classification may be summarized in Table I.

TABLE I

	categorical	probabilistic
precise	$x = s_j$	$y = p$
imprecise	$x \in S'$	$y \in P'$

If non-randomness is thought of as the limiting case of randomness, i.e. in terms of distribution using only probabilities either 0 or 1, the degree of generality of the above four cases can be indicated by the following graph, where the arrows denote the direction of specialization:

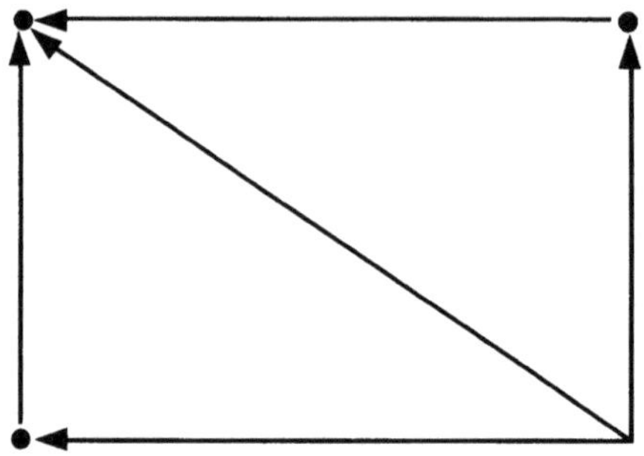

Next property of information to be taken into account is its reliability. Or, rather, this is a property of the source of information: a source of information is said to be reliable if all information emanating from it is true. Unreliable sources are characterized by the probabilities of different specific pieces information, conditional upon the actual state of Nature. Obviously, the value of information ought to depend the reliability of its source. We do not need to go here into this in a more detailed way, for we will assume that the source is reliable. It follows that all information is assumed to be true.

An information which is precise, categorial and emanating from a reliable source will be called 'perfect'. In what is going to follow we will restrict our attention to information of this type.

Information, as conceived in this paper, is easily seen to be of direct, or a *priori*, type. Another possible kind of information is that provided by the outcome of an experiment, such that the probability distribution on the set of all possible outcomes depends upon the state of Nature. This is usually called indirect, or *a posteriori*, information (or evidence). The use to which such information can be put has been the subject of mathematical statistics or, at least, of the school of statistical thought which originated with A. Wald.

Although the word 'information' is freely used in such a context, there have been but few attempts to give it a precise meaning (notable exceptions: Fisher, Lindley). As far as I know, it still remains an open problem to assess the pragmatic value of such information. A way to do this has been indicated by Wald, who introduced explicitly the cost of experimentation into the conceptual framework of mathematical statistics.

Indirect information will not be discussed here in any systematic way. I shall only remark that direct information emanating from an unreliable source can be shown to be a special case of indirect information.

II GENERAL REMARKS ON THE VALUE OF INFORMATION

Suppose now that a decision-maker, faced with the problem $\|u_{ij}\|$, is about to receive perfect information concerning S. What is this information worth to him?

Obviously, specific pieces of information of this kind are not, in general, equally 'good' (in the sense of good and bad news). To put it more precisely, of any two such pieces of information one will, in general, make it possible to obtain higher utility by a suitable choice of action than the other. But the decision-maker does not know in advance which particular information he will receive. For such knowledge would make the information redundant. Hence, the pragmatic value we are concerned with should be associated not with any particular statement of the form $x = s_j$, but, rather, with the set of all such statements, i.e. with the set of all possible informations of the given kind. Let us call this set E. In other words, we are concerned with assessing the value of a direct (and true) answer to the question: which state of Nature actually obtains? This way of looking at the problem can easily be generalized to cover information of any other type.

The value we try to define must, of course, depend upon the use the information is put to. This may be described generally by means of a function from E to A, i.e., by associating with each possible information an action to be chosen if this particular information is received. Any such function, denote it by d, will be called decision function,[2] in view of the obvious analogy with the notion introduced by A. Wald. The class of all decision functions, denoted by D, includes the set A of all actions, for any a_i belonging to A may be thought of as such d that $d(e) = a_i$ for all e in E. Receiving information is thus equivalent to extending of the set of actions available to the decision-maker.

The next step is to see what the decision-maker stands to gain when he utilizes the information according to a decision function d. This can be done as follows:

Any pair, $d \in D$ and $s \in S$, determines an outcome (in utility). For exactly one piece of (perfect) information is compatible with the state of Nature s, and this information induces, by d, the choice of action; the action chosen, together with s, determines the outcome. In the case of imprecise information the problem of outcome becomes more complex. Being now concerned with perfect information only, we need not go into the description of such a complex case. We may safely assume that obtaining information of given type involves a non-negative cost, say C, the unit of C being the same as that used in the description of the decision problem. This cost must be subtracted from the utility of all outcomes of any decision function (proper, i.e. belonging to the set $D - A$). The problem now arises of making use of one's information, with the action available previously, i.e., when no information was provided. An answer to the last question is equivalent to adopting a criterion of decision making. For any such criterion establishes an order on the set of actions available and, by the same token, on the D. Without such an ordering it would be impossible to compare different ways of utilizing information with the action available previously. Let us suppose that some such criterion has been adopted. We will then define the value of information as *the highest value of C, such that the optimal decision function* (*in the sense of this criterion*) *belongs to the set $D - A$*. In other words, the value of information is the highest price the decision-maker would be prepared to pay for it. It is assumed that he not prepared to buy information if, when utilizing it in best way he knows, he is worse off than without this information.

The meaning of the above definition will, perhaps, become clearer when we proceed to discuss it in terms of several possible criteria of decision-making. Meanwhile, let me stress again some properties of the above definition: (1) The value of information concerning S is always relative to a decision problem where the consequences of available actions depend on elements of S. (2) This value is associated with the set of all possible informations of a given type. (3) It is relative to the criterion of decision-making.

I am now going to recall briefly some proposed solutions of the decision problem. First there are some the criteria suggested for solving the problem when no *a priori* distribution on S is assumed (this is called decision-making under uncertainty). Each of them establishes an order on the set of actions: the maximal element (or elements) of this order is, by definition, optimal in the sense of the criterion of question. The following conditions define various meanings of optimality. I denote the optimal action by a_r.

Maximin: $\min_j u_{rj} = \max_i \min_j u_{ij}$;

Laplace: $\frac{1}{n}\sum_j u_{rj} = \max_i \frac{1}{n}\sum_j u_{rj}$;

Minimal loss: $\max_j w_{rj} = \min_i \max_j w_{ij}$,

where w_{ij} = the loss incurred when a_i chosen while s_j is equal to $\max_i u_{ij} - u_{ij}$.

Hurwicz:

$$\alpha \min_j u_{rj} + (1-\alpha)\max_j u_{rj} = \max_i \left[\alpha \min_j u_{ij} + (1-\alpha)\max_j u_{ij}\right].$$

where α, called sometimes the index of pessimism, is a parameter characterizing the decision maker; $0 \leq \alpha \leq 1$.

β-criterion[3]:

$$\beta \min_j u_{rj} + (1-\beta)\frac{1}{n}\sum_j u^{rj} = \max_i \left[\beta \min_j u_{rj} + (1-\beta)\frac{1}{n}\sum_j u_{ij}\right],$$

where β is a parameter analogous to α above; $0 < \beta < 1$.

The relative merits of the above criteria have been discussed by several authors, in terms of certain postulates of rationality. There is no need to recount this discussion here: any of above solutions (or any other reasonable solution, for that matter) will do in order to give specific meaning to the expression 'the value of (perfect) information, relative to a decision problem'.

Another approach consists in assuming a (subjective) *a priori* distribution on S, representing the judgment of the decision-maker as to the chances of different s_j's to happen. The distribution has to be subjective, for we have assumed that the actual state of Nature does *not* vary randomly. In such case, called 'decision-making under risk', the criterion is obviously that of maximizing expected utility. Denoting the distribution by $p = \langle p_1, \ldots, p_n \rangle$, we have the following condition of optimality for an action a_r.

$$\sum_j u_{rj}\, p_j = \max_j \sum_j u_{ij} p_j .$$

Before I proceed to define and discuss the value of perfect information in terms of the above criteria, let me make the following remark. In the case of perfect information, we may simplify our task considerably by reducing the set D to $A \cup \{d_0\}$, where d_0 is this decision function which associates with the information $x = s_j$ an action for which u_{ij} is maximal (if there are several such actions, any one of them will do). Obviously, d_0 dominates in the game–theoretic sense every decision function in $D - A$, i.e., it is at least as good as any d in $D - A$, under all states of Nature. In other words, d_0 is *the* best way of utilizing perfect information. Therefore, it is only d_0 we need to compare with the elements of A. Under the jth state of Nature, d_0 yields the utility equal to: $\max_i u_{ij} - C$.

Notice that if information is, say, categorical and imprecise, this shortcut is no longer available. For then a decision function dominating any other function in $D - A$ does not, in general, exist.

Having completed this, somewhat lengthy, introduction, I am now going to consider the value of perfect information in the light of the above criteria in turn.

III MAXIMIN

The decision function d_0 is maximin-optimal with respect to all actions A if the following condition is satisfied:

$$\min_j \left(\max_i u_{ij} - C \right) \geq \max_i \min_j u_{ij} . \tag{1}$$

(The left-hand side of the inequality is he maximin index of d_0, the right-hand side is the highest maximin index for actions in A.) Hence

$$\min_j \max_i u_{ij} - \max_i \min_j u_{ij} \geq C . \tag{2}$$

The expression on the left-hand side is familiar from the theory of games. It is known to be non-negative, hence (2) is consistent with the assumption that $C \geq 0$. The maximal value of C satisfying (2) is

$$C_w = \min_j \max_i u_{rj} - \max_i \min_j u_{rj} . \tag{3}$$

The above expression defines the value C_w of perfect information, relative to the maximin criterion of decision making.[4] Obviously,

$$0 \leq C_w \leq \max_i \max_j u_{ij} - \min_i \min_j u_{ij}.^5 \tag{4}$$

A sufficient and necessary condition for this value to be minimal, i.e. equal to zero, is the existence of a saddle point in the decision matrix. If a decision maker is committed to maximin and faced with a problem of such a type, perfect information has for him no value at all. This seems somewhat paradoxical. It is a counterpart of another peculiar property of the maximin criterion. It has been shown elsewhere[6] that if the decision matrix has a saddle point, no decision function based on the outcome of an experiment is better, in the sense of maximin, than the action with the saddle point. In other words, experimental evidence, however compelling, is perfectly useless to a decision-maker committed to maximin if his problem has a saddle point. I consider this to constitute a rather strong objection to the use of the maximin criterion as a guide to decision under uncertainty.

What are the decision problems for which C_w, as defined by (3), attains maximum? Strictly speaking, there is, of course, no upper limit to the value of C_w, for it depends on the choice of unit of utility in the decision problem. This last choice being arbitrary, C_w may be as high as we like. Supposing, however, that the unit is fixed, we may look for a structure of the decision problem which would make the value of perfect information maximal. Abbreviating the overall maximum (minimum) by

$$\begin{aligned} \max &= \max_i \max_j u_{ij}, \\ \min &= \min_i \min_j u_{ij}, \end{aligned} \tag{5}$$

we have as a sufficient and necessary condition for C_w to be maximal:

$$\begin{aligned} &\max_i u_{ij} = \max \text{ for all } j, \\ &\min_j u_{ij} = \min \text{ for all } i. \end{aligned} \tag{6}$$

The above condition amounts to postulating that all column-maxima and all row-minima be equal. The maximin value of perfect information is then:

$$C_w = \max - \min. \tag{7}$$

A special case is the identity matrix (having 1 along the main diagonal and 0 elsewhere). Such a matrix may be interpreted as representing a 'purely cognitive' problem, i.e. that of determining the actual state of Nature: the value of success is 1, the value of failure is 0. For such a problem, the maximin value of information is maximal.

IV LAPLACE

According to the argument used in the case of maximin, the best decision function based on perfect information is Laplace optimal with respect to A iff

$$\frac{1}{n}\sum_j\left(\max_i u_{rj} - C \geq \max_i \frac{1}{n}\sum_j u_{rj}\right). \tag{8}$$

The Laplace value of perfect information, say C_1, is thus

$$C_L = \frac{1}{n}\sum_j \max_i u_{ij} - \max_i \frac{1}{n}\sum_j u_{ij}, \tag{9}$$

or, in words, the difference between the average of column-maxima and the maximum of row averages. This difference is obviously non-negative, and it is equal to zero iff there exist an action, say a_r, which dominates all other actions:

$$u_{rj} \geq u_{ij} \quad \text{for all } i \text{ and } j. \tag{10}$$

This accords with intuition. Clearly, if in a decision problem an action is at least as good as any other action, under all states of Nature, no information concerning S is needed. And it seems reasonable that information has no value only in such a case.

In a finite decision problem, the existence of dominating action implies the existence of a saddle point. Therefore, if the Laplace value of information is equal to zero, so is its maximin value.

What about the maximum of C_L? Under the assumption that $m \geq n$, the following holds. The Laplace value of information is maximal iff all the column-maxima are equal and every row contains at most one such maximum, all other entries being equal to the overall minimum. Then:

$$C_L = (\max - \min)\left(1 - \frac{1}{n}\right). \tag{11}$$

The maximum value of information in the sense of Laplace is thus an increasing function of n. A particular case of the decision problem with such a structure is again, the identity matrix for which $C_L = 1 - 1/n$.

V MINIMAX LOSS

The loss associated with d_0 is

$$\max_i u_{ij} - \left(\max_i u_{ij} - C\right) = C \text{ for every } j. \tag{12}$$

We thus have as the condition of optimality for d_0 :

$$C \leq \min_i \max_j\left(\min_i u_{ij} - u_{ij}\right). \tag{13}$$

Hence, the minimax loss value of perfect information, say C_S, is

$$C_s = \min_i \max_j \left(\max_i u_{ij} - u_{ij} \right). \tag{14}$$

The decision problem for which $C_S = 0$ is such that for some r

$$\max_i u_{ij} - u_{rj} = 0 \quad \text{for all } j, \tag{15}$$

which amounts to saying that there exists an action a_r dominating all other actions.

On the other hand, C_S attains maximum iff for each action the maximal loss attains the highest value possible, i.e. max-min. This, in turn, happens iff for each action a_i there exist such a state s_j that:

$$u_{ij} = \min \quad \text{and for some } k\colon u_{ik} = \max. \tag{16}$$

The value of perfect information is then equal to max-min. The identity matrix satisfies the above condition.

It is not difficult to see that the condition for maximizing C_L implies that for maximizing C_S.

VI HURWICZ

The condition of optimality for d_0 is

$$\alpha\left(\min_j \max_i u_{ij} - C\right) + (1-\alpha)\left(\max_j \max_i u_{ij} - C\right) \geq \max_i \left[\alpha \min_j u_{ij} + (1-\alpha) \max_i u_{ij}\right]. \tag{17}$$

Denoting the Hurwicz optimal action by a_r, we have from (17):

$$\alpha\left(\min_j \max_i u_{ij} - \min_j u_{rj}\right) + (1-\alpha)\left(\max_j \max_i u_{ij} - \max_j u_{rj}\right) \geq C \tag{18}$$

and the left-hand side is easily seen to be non-negative, complying with the condition that $C \geq 0$. The Hurwicz α-value of perfect information is

$$C_\alpha = \alpha \min_j \max_i u_{ij} + (1-\alpha) \max_j \max_i u_{ij} - \max_i \left[\alpha \min_j u_{ij} + (1-\alpha) \max_j u_{ij}\right]. \tag{19}$$

Setting $\alpha = 0$ we have $C_\alpha = 0$, whatever the decision problem. Hence, a radical optimist, in the sense of the Hurwicz parameter, will attach no value to perfect information, no matter what his decision problem is, which seems rather absurd. C_α being continuous with respect to α, the more we approach radical optimism, the closer we get to this absurdity. The Hurwicz value of information $C_\alpha = 0$ for all α iff the decision problem has a saddle point and one of the entries in the row containing the saddle point is equal to the overall maximum.

PROOF: Sufficiency is obvious. Necessity:

(1) Assume there is no saddle point, i.e., $\max_i \min_j u_{ij} < \min_j \max_i u_{ij}$.
Then for $\alpha = 1$, $C_\alpha > 0$.

(2) Assume there is a (unique) saddle point, say u_{rs}, but $\max u_{rj} < \max = u_{wt}$, where $w \neq r$. Then for some $\alpha > 0$ the action a_w is α – optimal in A, hence α – better than a_r, i.e.

$$\alpha\left(u_{rz} - \min_j u_{wj}\right) + (1-\alpha)\left(\max_j u_{rj} - u_{wt}\right) < 0 \tag{20}$$

This follows from the fact that

$$u_{rz} - \min_j u_{wj} > 0 \quad \text{and} \quad \max_j u_{rj} - u_{wt} < 0. \tag{21}$$

By a suitable choice of $\alpha > 0$ the left-hand side of (20) can therefore be made negative. But the same choice of α makes C_α positive:

$$C_\alpha = \alpha\left(\min_j \max_i u_{ij} - \min_j u_{wj}\right) + (1-\alpha)\left(\max_j \max_i u_{ij} - u_{wt}\right) > 0, \tag{22}$$

because $\min_j \max_i u_{ij} > \min_j u_{wj}$ while the second component of the sum vanishes. Therefore, if the condition is not satisfied, $C_\alpha > 0$ for some α. This completes the proof. It is easily verified that C_α is maximal for all α iff the column-maxima are equal (being thus equal to max) and all those rows which contain max, have also min. Again, the identity matrix exemplifies such a case. The Hurwicz α-value of perfect information is then:

$$C_\alpha = \max - [\alpha \min + (1-\alpha)\max] = \alpha(\max - \min), \tag{23}$$

the maximum α-value of perfect information is thus an increasing function of the index of pessimism.

VII β-CRITERION

By the same argument as that used above, we get as the β-value of perfect information

$$C_\beta = \beta \min_j \max_i u_{ij} + (1-\beta)\frac{1}{n}\sum_j \max_i u_{ij} - \max_i\left[\beta \min_j u_{ij} + (1-\beta)\frac{1}{n}\sum_j u_{ij}\right]. \tag{24}$$

It is easily seen that for all values of β, C_β is non-negative. $C_\beta = 0$ for all iff there exist a dominating action.

PROOF: Sufficiency is obvious, in view of the fact the existence of a dominating action implies the existence of saddle point. Necessity: suppose no dominating action exists. Let us denote the optimal action in A by a_r. Then:

$$C_\beta = \beta\left(\max_j \max_i u_{ij} - \min_j u_{rj}\right) + (1-\beta)\frac{1}{n}\sum_j\left(\max_i u_{ij} - u_{rj}\right). \tag{25}$$

The first component is non-egative, the second one is positive (because, by assumption, a_r is not a dominating action). Hence, $C_\beta > 0$ for any β, which proves the necessity. As to the maximum - value of information (assuming, as in the case of Laplace, $m \geq n$): C_β = maximum for all iff (1) $\max_i u_{ij}$ = max for all j; (2) in every row, max appears at most once, all other entries being equal to min. Then:

$$C_\beta = \max - \beta \min - (1-\beta)\frac{1}{n}[(n-1)\min + \max]$$

$$= (\max - \min)\left(1 - \frac{1-\beta}{n}\right). \tag{26}$$

The condition for maximizing (for all β) the β-value of perfect information turns out to be the same as that for Laplace value.

VIII MAXIMIZATION OF EXPECTED UTILITY

Now, it is assumed that decision-maker acts according to his expectation concerning S, represented by a subjective probability distribution on S, say $p = \langle p_1, \ldots, p_n \rangle$.
Expected utility of the action optimal in A is equal to

$$\max_i \sum_j u_{ij} p_j. \tag{27}$$

The expected utility of d_0 is equal to

$$\sum_j\left(\max_i u_{ij} - C\right)p_j = \sum_j \max_i u_{ij} p_j - C. \tag{28}$$

Therefore, d_0 is optimal with respect to the action in A iff

$$\sum_j \max_i u_{ij} p_j - \max_i \sum_j u_{ij} p_j \geq C. \tag{29}$$

Hence, the value C_p of perfect information, relative to the distribution p, is

$$C_p = \sum_j \max_i u_{ij} p_j - \max_i \sum_j u_{ij} p_j, \tag{30}$$

or, equivalently,

$$C_p = \min_i \sum_j w_{ij} p_j, \tag{31}$$

where w_{ij} is the loss associated with a_i and sj, defined as $w_{ij} = \max_i u_{ij} - u_{ij}$. C_p thus turns out to be the minimum average loss, with respect to the distribution p. The transition from (30) to (31) is possible in view of the identity:

$$\sum_j \max_i u_{ij} p_j - \max_i \sum_j u_{ij} p_j = \min_i \sum_j (\max_i u_{ij} - u_{rj}) p_j. \tag{32}$$

This can be proved in the following way. The left-hand side of (32) is equal to

$$\sum_j \max_i u_{ij} p_j - \sum_j u_{rj} p_j = \sum_j \left(\max_i u_{ij} - u_{rj} \right) p_j. \tag{33}$$

iff a_r satisfies the condition

$$\sum_j u_{rj} p_j \geq \sum_j u_{ij} p_j \quad \text{for all } i. \tag{34}$$

The right-hand side of (32) is equal to

$$\sum_j \left(\max_i u_{ij} - u_{rj} \right) p_j \tag{35}$$

iff a_r satisfies the condition

$$\sum_j \left(\max_i u_{ij} - u_{rj} \right) p_j \leq \sum_j \left(\max_i u_{ij} - u_{rj} \right) p_j \quad \text{for all } i. \tag{36}$$

Therefore, in order to prove (32), it is enough to prove that the conditions (34) and (36) are equivalent. But their equivalence is obvious. Hence (32) is valid.

Some theorems show the relation between the decision problem $\|u_{ij}\|$, the distribution p on S, and the value of perfect information, relative to both $\|u_{ij}\|$ and p. (Obviously, $0 \leq C_p \geq \max - \min$.)

$$C_p = 0 \tag{37}$$

for all p iff there exists a dominating action.

PROOF: Sufficiency is obvious. Necessity: if $C_p = 0$ holds for all p it holds, in particular, for a p such that $p_j > 0$ for all j. Denote by a_r the optimal action, with respect to such a distribution p. Then:

$$C_p = \sum_j \left(\max_i u_{ij} - u_{rj} \right) p_j \tag{38}$$

is equal to 0 iff $\max_{iu} ij = u_{rj}$ for all j, which shows that a_r is dominating in A.

Next, something may be said about the distribution minimizing C_p with respect to all decision problems.

$$C_p = 0 \tag{39}$$

for all decision iff there exists such k that $p_k = 1$. This, again, accords with intuition. If the decision maker is subjectively certain that a particular state of Nature will occur, information has, for him, no value, whatever his decision problem depending on S. And only in such a case will this happen.

PROOF: Sufficiency: in view of $p_j = 0$ for all $j \neq k$, we have

$$C_p = \max_i u_{ik} p_k - \max_i (u_{ik} p_k) = 0. \tag{40}$$

Necessity: suppose (the argument can easily be generalized) that $p_j > 0$ for exactly two states, say s_k and s_l. It is then easy to construct a problem for which $C_p > 0$. Let it consist of two actions, a_1 and a_2, such that $u_{1k} > u_{2k}$ and $u_1 < u_{2l}$. Then

$$\max_i u_{ik} p_k + \max_i u_{il} p_l > u_{ik} p_k + u_{il} p_l \quad \text{for } i = 1, 2. \tag{41}$$

Hence $C_p > 0$, which proves the necessity.

Let us now look at the decision problem which maximizes C for all p. The condition which held for the Laplacean distribution turns out to hold for any distribution p. If (for simplicity's sake) we restrict our attention to such decision problems that $m \geq n$, a sufficient and necessary condition for

$$C_p = \text{maximum for all } p \tag{42}$$

is: all the column-maxima are equal to the overall maximum: in every row this maximum appears at most once, all the other entries being equal to the overall minimum.

If this condition is satisfied, the value of perfect information, with respect to p, is

$$C_p = (\min - \max)\left(1 - \max_j p_j\right). \tag{43}$$

The proof is easy but somewhat tedious. Therefore it will be omitted in this exposition. Another question that may be asked is: given a decision problem $\|u_{ij}\|$, what is the distribution p which maximizes p-value of perfect information, with respect to this problem? In view of (31), the distribution maximizing C_p is that maximizing the loss, induced by the original matrix of utility. Therefore, the distribution in question is found by solving the game defined by the loss matrix. (Strictly speaking, this answer is valid if the set A of action is extended to the class of all mixed actions.)

IX SUMMARY

Perfect information has been defined as a categorical and precise (direct) answer to the question: which element of the set of states S actually obtains? It has also been assumed that the answer emanates from a reliable source and is, therefore, true.

The problem was to define the value of perfect information, relative to a decision problem where the consequences of available actions are determined by the elements of *S*.

The value in question was identified with the highest price (in utility) to be paid for the information, compatible with the condition that the best use made of perfect information is at least as good as any action previously available. The expression 'the best use' is here interpreted as 'dominating any other use in game-theoretical sense'. The expression 'at least as good' is given several alternative meanings, according to different criteria of decision making.

The resulting variants of the definition have been investigated. The typical question was, what type of decision problem minimizes (maximizes) the value, in the given sense, of perfect information. No answer to this question was found to be common to all definitions of value, induced by different criteria. In most cases, information has minimal (i.e., zero) value iff there is a dominating action. A notable exception is the maximin case, where the existence of a saddle point is not only necessary but also sufficient. In all cases, the identity matrix (interpreted as a purely cognitive decision problem) is an instance of the typical problem maximizing the value of perfect information.

The above line of argument can be applied to at least some other types of information, mentioned in the introduction. The resulting definitions would be essentially analogous in character. In the present paper the attention was restricted to perfect information, which obviously is the simplest case.

NOTES

[1] Colin Cherry, *On Human Communication: A Review, a Survey, and a Criticism*, New York 1957, p. 243.

[2] Still more generally, a decision function could be defined as that from *E* to the set of mixed actions, i.e., to the class of all probability distributions on *A*. We will, however, restrict out argument to the simpler case.

[3] Cf. K. Szaniawski, 'Some Remarks Concerning the Criterion of Rational Decision-Making', *Studia Logica* 9 (1960) 221–239.

[4] Let us note, in passing, that this value is invariant under the change of origin on the utility scale. This remark applies to all variants of the proposes definition, i.e., it remains valid whatever criterion is adopted.

[5] This, again, is valid in all cases considered.

[6] See note 3 above.

17
QUESTIONS AND THEIR PRAGMATIC VALUE

Let S be a set. The elements of S will be called 'states (of the world)'. They are assumed to be exhaustive and pairwise disjoint (the formal expression of this assumption is easily obtained if the elements of S are interpreted as sentences).

By 'question concerning S' we shall mean the ordered pair $Q = \langle X, \{p_s\}_{s \in S} \rangle$, where X is an exhaustive set of pairwise disjoint events and p_s is a family of probability distribution on X, parametric with respect to S. In other words, a probability distribution p_s on X is associated with s, for each $s \in S$. The elements of X will be called 'answers'.

The above definition of 'question' seems to cover all the essential uses of the notion. Thus, for instance, Q is a categorical question concerning $\underline{S}$ if all the probability distributions p_s are of zero-one type.

If Q is categorical, two cases are possible. (1) For each x, there is exactly one s such that $p_s(x) = 1$. This establishes a one-one correspondence between X and S. An answer x can then be interpreted as the statement that the actual state of world is this s which makes x certain. (2) For some x, there is more than one s such that $p_s(x) = 1$. In this case X defines a partition of S, the elements of which are: $S_x = \{s\colon p_s(\underline{x}) = 1\}$, for $x \in X$. An answer x is then interpreted as the statement that the actual state of the world belongs to S_x. The remaining case of some being such that $p_s(\underline{x}) = 0$, for all $s \in S$, can be reduced to either (1) or (2), in view of the assumption that the set S is exhaustive; for such answers are impossible under all states of the world and, therefore, can be eliminated from the set X.

A question can be 'nearly categorical' if the distributions p_s approximate the zero-one type; i.e., if they are unimodal, the modal probability being sufficiently high. It is then natural to interpret the most probable answer, under the state s, as the statement that the actual state of the world is s (that it belongs to S_x, respectively). Now, however, the answer may be false, in contrast to the categorical question. The degree of reliability of the answer is determined by values of the modal probabilities. The above considerations show that semantic interpretation of answers (i.e., of the elements of X), if any, is entirely determined by the relation between S and X, i.e., by the character of the probability distributions p_s. Indeed, the semantic interpretation is secondary to the probabilistic relations between S and X. In some cases such an interpretation seems formally inessential. Consider, for example, the above-mentioned case of unreliable solution of Q. We can interpret X in this way if for each x there is an s such that $p_s(x)$ is 'sufficiently high'. It is difficult to see how this limitation imposed upon the distributions p_s could be exploited in a theory of questions. Hence, it seems advisable to consider a general case, i.e., to impose no limitations on the distributions p_s, rather than single out 'unreliable solutions of Q' as a special case.

The pragmatic value of a question Q concerning S can be defined relative to a decision problem the outcomes of which depend on S. Given such a problem $U = \langle A, S, u\rangle$, where A is the set of actions, and u is a utility function defined on the Certasian product $A \times S$, any criterion of decision-making (maximization of expected utility included) generates a valuation on the question. The valuation is based on the fact that the choice of action can, in principle, be improved (in the sense of the criterion adopted) if it is made dependent on x.

Thus, in the general case i.e., when no limitations are imposed on the character of probability distributions p_s, the value of the question Q, relative to a decision problem and (for instance) an *a priori* probability distribution p on S is determined as follows.

Any use that can be made of having the question Q answered is represented by a decision function d from X to A. The set D of all d is an extension of A, in view of the fact that an action a in A is equivalent to a d such that $d(x) = a$ for all x in X. The role played by Q consists therefore in providing the decision-maker with some additional ways of behaving. The profit expected from such an extension is defined by the difference between the value of the optimal d in D and that of the optimal a in A. In the case under consideration, i.e., when the probability distribution p on S is given, this difference can easily shown to be

$$C(Q;\, U, p) = \sum_x \max_a v(a, x) - \max_a \sum_x v(a, x),$$

where

$$v(a, x) =_{df} \sum_s u(a, s)p_s(x)p(s).$$

$C(Q;\, U, p)$ is read as: the value of the question Q, relative to the decision problem U and the *a priori* probability distribution p. So defined, the value of a question is, of course, non-negative: if in a given case the solution of Q is useless, relative to U and p, the optimal d is identical with optimal a, which makes C equal to zero.

If the *a priori* distribution is not assumed to be given, a criterion of decision making (such as for example, maximin or minimax loss) plays the role of maximization of expected utility. Several problems arise in connection with the maximal and minimal value of $C(Q;\, U, p)$. A number of them have been solved. The notion of the value of a question also has a bearing on a valueless choice, relative to a non-zero trivial decision problem (i.e., such a problem in which no dominating action exists.)

18
TWO CONCEPTS OF INFORMATION*

0 INTRODUCTORY REMARKS

Among different meanings of the word 'information', there is at least one which is closely connected with decision-making. It is usually referred to as 'pragmatic information',[1] the adjective indicating that what we have in mind is information for some purpose. There has been, to my knowledge, no attempts to explore systematically the notion of pragmatic information. Elsewhere I tried to define the concept ([7], [8]) and to investigate some of its properties for the special case of categorical, i.e. non-probabilistic, information ([6], [7]). The main purpose of the present paper is to compare the concept of pragmatic information with the classical, entropy-based notion, due to Shannon. In order to do this I shall have to assume that the joint probability distribution of a two-dimensional random variable is given, for this is what the statistical, i.e. Shannon's concept of information presupposes. I shall also make another special assumption to be explained below. Having made the relation between the two concepts apparent, I will indicate some consequences of lifting of the two restrictions.

1 PRAGMATIC INFORMATION, PROBABILITY FULLY SPECIFIED

Let $\langle X, S\rangle$ be a two-dimensional random variable with the joint probability distribution function

$$p(x, s) = Pr\{x = x \wedge sS = s\} \tag{1}$$

defined for all $x \in X$ and $s \in S$, where X and S are the sets of values of the variables X and S, respectively. For the sake of simplicity, I shall assume that all the sets spoken of here are finite.

The absolute and conditional probabilities, obtainable from $p\,(x, s)$, will be denoted by

$$p(s) = \sum_x p(x, s), \tag{2}$$

$$p(x \mid s) = p(x, s)/p(s) \tag{3}$$

and, analogously, by $p(x)$, $p(s \mid x)$.

Let us further assume that a decision problem $U = \langle A, S, u \rangle$ is given, where A = the set of possible actions, S = the set of states the world, u = utility, i.e. a real function defined on $A \times S$ and representing valuation by the decision maker of all the combinations: action a taken, state s occurs ($a \in A$, $s \in S$). The valuation is extended over the probability mixtures of the elements of $A \times S$, according to the principle (von Neumann and Morgenstern) that the expected value of utility is utility.

The connection between $\langle X, S \rangle$ and $\langle A, S, u \rangle$ is easily seen to consist in the fact the set S of states of the world is identical with the set of values of the random variable S. Thus the actual value of the variable S determines the consequences of actions in A. Obviously, if the purpose is to maximize u by a suitable choice of action from the set A, the decision could, in general, be improved by providing the decision-maker with some information concerning the actual value of S.

This is where the variable X comes in. The knowledge of the value of X reduces, through $p(x, s)$, the uncertainty concerning S (on the average, of course, and if we except certain special cases which will be discussed later). Hence the possibility of improving the decision. Pragmatic information is precisely the measure of this improvement.

To make use of the information provided by X in decision making means to make the choice of action depend on the value X takes. This is achieved by a decision function (the term is due to A. Wald), i.e. by a function d from X to A. The interpretation of d is as follows: $d(x) = a$ means that the decision-maker is committed to take action a in the case of $X = x$. Of course, there are many such functions (the set of all d will be denoted by D). However, in the case under consideration, it is fairly easy to determine which of them is optimal. Clearly, the best thing to do is to associate with any given x such an action which maximizes expected utility under the probability distribution $p(s \mid x)$. Formally, an optimal decision function is defined by the following formula, valid for all x and a:

$$d_{opt}(x) = a \Rightarrow \sum_s u(a, s)p(s \mid x) = \max_a \sum_s u(a, s)p(s \mid x). \tag{4}$$

And by averaging this maximal utility over the set X we obtain

$$u(d_{opt}) = \sum_x \max_a \sum_s u(a, s)p(x, s), \tag{5}$$

which is the utility associated with the optimal decision function.[2]

On the other hand, if information provided by X is not available, the best course of action is for the decision maker to choose this a which maximizes expected utility under the probability distribution $p(s)$. The utility associated with such a_{opt} is:[3]

$$u(a_{opt}) = \max_a \sum_s u(a, s)p(s) \tag{6}$$

The increase (if any) of utility, due to the variable X, is equal to the difference between $u(d_{opt})$ and $u(a_{opt})$. According to the previous remarks, this measures pragmatic information.

Pragmatic information will be denoted by $C(X, S; U)$ to bring out the fact that it depends on the decision problem U.

$$C(X, S; U) = u(d_{opt}) - u(a_{opt}). \tag{7}$$

Substituting (5) and (6) in (7) we get

$$C(X, s; U) = \sum_x \max_s \sum_s u(a, s)p(x, s) - \max_a \sum_s u(a, s)p(s). \tag{8}$$

If we take (2) into account and introduce the abbreviation

$$v(a, x) = \sum_s u(a, s)p(x, s), \tag{9}$$

C can be represented more symmetrically as

$$C(X, S; U) = \sum_x \left(\max_a v(a, x) - \max_a \sum_x v(a, x)\right), \tag{10}$$

which, of course, reduces to

$$C(X, S; U) = \min_a \sum_x \left(\max_a v(a, x) - v(a, x)\right). \tag{11}$$

As all elements in the above sum are, by the definition of maximum, non-negative, the sum itself is also non-negative and it attains its lowest value, i.e. zero, iff

$$\bigvee_s \bigwedge_x \left(v(a, x) = \max_a v(a, x)\right) \tag{12}$$

In view of

$$v(a, x) = \max_a v(a, x) \Leftrightarrow \sum_s u(a, s)p(s \mid x) = \max_s \sum_s u(a, s)p(s \mid x). \tag{13}$$

the meaning of the condition (12) is fairly obvious. (12) in fact states that there exists such an action in A which, for *all* x, maximizes expected utility under the conditional probability distribution $p(s \mid x)$. In other words, there exists an action, optimal absolutely, i.e. optimal whichever value the variable X assumes.

It is hardly surprising that (12) is a sufficient and necessary condition for the pragmatic information to be zero. If an action is invariably adopted, no matter which statement of the form $X = x$ happens to be true, then the solution of the decision problem U does not depended on the information. Following the terminological suggestion by Mr G. Lissowski, I shall say in such case that S is independent of X according to U.

Whether the condition (12) holds or not depends both on the decision problem U and on the joint probability distribution $p(x, s)$. This may be instantiated by two special cases. If U is such that an action dominating in utility all other actions exists. i.e.

$$\bigvee_{a*} \bigwedge_a \bigvee_s (u(a^*, s) \geq u(a, s)), \tag{14}$$

then (12) holds and $C(X, S; U) = 0$ for all probability distribution functions $p(x, s)$. Also, if the variables X and S are stochastically independent, i.e.

$$\bigwedge_x \bigvee_s (p(x, s) = p(x)p(s)). \tag{15}$$

then S is independent of X according to any U, and $C(X, S; U) = 0$ for all U. This is easily seen. Assuming that (15) holds, we have

$$C(X, S; U) = \sum_x \max_a \sum_s u(a, s)p(x)p(s) - \max_a \sum_s u(a, s)p(s) = 0. \tag{16}$$

2 STATISTICAL INFORMATION

The Shannon measure, often called statistical information, is defined as

$$I(X, S) = H(S) - H(S \mid X). \tag{17}$$

In the above expression $H(S)$ is the entropy of the variable S, i.e. the expected value of the logarithm of its probability distribution function (with the minus sign, to make it non-negative):

$$h(s) = -\sum_s p(s) \log p(s) \tag{18}$$

$H(S \mid X)$ is conditional entropy $H(S \mid X) = x$, averaged with respect to X:

$$H(S \mid X) = -\sum_x p(x) \sum_s p(s \mid x) \log p(s \mid x). \tag{19}$$

Entropy being a measure of uncertainty associated with a random variable, it easily seen that $I(X, S)$ measures the average decrease of uncertainty concerning S, due to the information brought by the variable X.

The philosophical importance of type of information concept has been stressed by Hintikka in [3]. This paper contains also an admirable discussion of varieties of such an approach.

3 STATISTICAL AND PRAGMATIC INFORMATION: A COMPARISON

A look at (7) and (17) is enough to notice certain similarity between the two concepts of information. In both cases, the situation where knowledge of the value assumed by the variable X is available, is compared with the case where it is not. And the 'gain' made possible by the knowledge of X is measured, either in utility (for C) or in decrease of uncertainty (for I). The similarity could be made even more explicit if loss, i.e. negative utility, were used instead of utility in the definition of C.

Further comparison is made difficult by the fact that pragmatic information depends on the decision problem U, whereas statistical information does not. Perhaps the simplest way to clarify this is to assume the decision problem to be fixed and carry out the comparison, relative to this special kind of U.

Neither is the choice of a suitable U difficult to make. The decision problem, let us denote it by E, will be defined as follows. The set A of actions is assumed to be identical with S, and the utility function will be binary:

$$u(a, s) = \begin{cases} 1 & \text{if } a = s, \\ 0 & \text{if } a \neq s. \end{cases} \tag{20}$$

The interpretation of E is straightforward. The decision in this case consists in choosing one of the states of the world. If the state chosen coincides with the state which actually obtains, the outcome is 'success' and is evaluated as 1; otherwise it is 'failure' and is equal to 0. Such a problem (guessing the actual state of the world) may be called epistemic, following Hempel who introduced the expression 'epistemic utility'.[4]

When U has the special form of E, the definition of C simplifies considerably. All but one elements of the sum over S vanish and utility no longer appears explicity in the expression. As the result we have

$$\begin{aligned} C(X, S; E) &= \sum_{x-} \sum_{s} p(x, s) - \max_{s} p(s) \\ &= \sum_{x} \max_{s} p(x, s) - \max_{s} \sum_{x} p(x, s) \\ &= \min_{s} \sum_{x} \left(\max_{s} p(x, s) - p(x, s) \right) \end{aligned} \tag{21}$$

which depends exclusively on the joint distribution function $p(x, s)$, just as $I(X, S)$ does. Let us now recall some well-known properties of statistical information. It satisfies the following inequalities:

$$0 \leq I(X, S) \leq H(S) \leq \log n, \tag{22}$$

where n is the number of elements in the set S.

As a counterpart of (22), we have, for pragmatic information relativized to E:

$$0 \leq C(X, S; E) \leq 1 - \max_{s} p(s) \leq 1 - n^{-1}. \tag{23}$$

Moreover, both C and I reach their respective critical values under similar conditions. Let us examine this in more detail.

A sufficient condition for the two expression to become equal to zero is the stochastic independence of the variables X and S:

$$\text{If (15) holds, then } I(X, S) = C(X, S; E) = 0. \tag{24}$$

The results for I are well known (in fact, if implication is replaced by equivalence, it may be said to define I as the only function which satisfies it); its validity for C has been shown above.

It may be remarked in passing that the necessary and sufficient condition for $C(X, S; E)$ to be zero, i.e. the special case of (12), is

$$\bigvee_{s*} \bigwedge_{x}\left(p(x, s^*) = \max_{s} p(x, s)\right). \tag{25}$$

In other words, $C(X, S, E) = 0$ iff there exist a value of S which is most probable for all values of X.

The second critical point is defined by the following condition on $p(x, s)$:

$$\bigwedge_{x} \bigvee_{s}(p(s \mid x) = 1). \tag{26}$$

The condition states that if we know what value X assumes, we have the value of S uniquely determined. As this obviously represents the best possible case, such information will be called perfect. Now, if pragmatic or statistical information is perfect, it attains its highest value:

$$\text{If (26) then } I(X, S) = H(S) \text{ and } C(X, S; E) = 1 - \max_{s} p(s). \tag{27}$$

Both $H(S)$ and $1 - \max_s p(s)$ depend entirely on the absolute probability distribution of S. And both expression reach their respective maximal values if the probability distribution is uniform:

$$\bigwedge_{s}(p(s) = n^{-1}), \tag{28}$$

$$\text{If (26) and (28) then } \quad I(X, S) = \log n \text{ and } C(X, S; E) = 1 - n^{-1}. \tag{29}$$

The last expressions, $\log n 1 - n^{-1}$ are strictly increasing functions of n.

It thus turns out that pragmatic information, when relativized to the epistemic problem, varies in essentially the same way as stastical information. The formulae (15), (26) and (28) define sufficient conditions for the critical values of the two expressions.

The parallelism is not surprising. Statistical information consists in the reduction of uncertainty concerning the value of S; while pragmatic information with respect to the epistemic problem consists in improving the prediction of the value of S. Clearly, the probabilistic relation between the variables X and S must have the same kind of influence on the numerical value of both types of information.

Going a step further, one might suspect that under suitable assumptions concerning the form of utility functions, the two concepts can be made to coincide exactly, up to linear transformation. This possibility will not be explored here.

In terms of the present discussion, there are, of course, noticeable differences between the two concepts. To mention a few of them, sufficient *and* necessary conditions for information to become minimum are different: in the case of I it is a stochastic independence, and in the case of C it is an independence according to the epistemic problem. Also $I(X, S)$ is symmetric in the two variables, while $C(X, S; E)$ is not (which follows from the fact that the epistemic problem based on S is, in general, not identical with that based on X, for the two sets are not necessarily equipotent). There are, undoubtedly, several other discrepancies, not inconsistent with the essential similarity.

4 PRAGMATIC INFORMATION, PROBABILITY PARTLY DETERMINED

It has been assumed above that the probabilistic relation between he variables X and S is given in the form of the joint probability distribution function $p(x, s)$. Without this assumption it would have been impossible to compare pragmatic information with the statistical one.

Such an approach, however, is subject to the well-known criticism that it is unrealistic, unless one is prepared to adopt the subjectivistic view on probability. It is argued that in the expression

$$p(x, s) = p(x \mid s)p(s), \tag{30}$$

only the first factor is known (for all x and s), while the so-called *a priori* probabilities $p(s)$ are usually inaccessible and sometimes do not even make any sense.

It is not my purpose here to take sides on the much debated issue whether the so-called Bayesian approach (consisting in the use of *a priori* probabilities) is adequate or not. The number of serious problems involved in this classical controversy makes it impossible to state one's own view in a few sentences. Besides, it may be doubted that a definitive solution, in any sense of the word, is possible at all. The only point I want to make is that as long as differing views exist, it would be a sound policy not to prejudge the issue in the very definitions of probability-based concepts. Pragmatic information is a concept which lends itself to such treatment: it can be stretched to suit people of different philosophical persuasions on the matter of *a priori* probabilities.

Let us, therefore, replace the assumption of joint probability distributions by the following one. For each $s \in S$, a probability distribution $p(x \mid s)$ over X is given. The distribution function $p(x \mid s)$ may be interpreted either as genuine conditional distribution, or (if one does not want to assume that S varies randomly) as a parametric distribution with respect to s. Pragmatic information can then be defined in the following way.

For the decision function d and any s, the utility of adopting d when s is the actual state of he world is computed as

$$u(d, s) = \sum_{x} u(d(x), s)p(x \mid s). \tag{31}$$

In this way, the original decision problem $\langle A, S, u\rangle$ becomes extended to $\langle D, S, u\rangle$. I use the word 'extended' because the relation $A \subset D$ may be assumed to hold, considering that any $a \in A$ is identical with a decision function associating this a with all $x \in X$. The gain in utility due to such an extension is now identified with pragmatic information. The problem arises, how to assess the gain in question.

To be able to do this, we must have a means of comparing the elements of A with the elements of D-A, i.e. with non-trivial decision functions. To calculate the averages over S of $u(d, s)$ and $u(a, s)$ would be the obvious way of doing it if we assumed the *a priori* distribution $p(s)$ as known. This is, in fact, what we did previously: the optimal d (in terms of average utility) was then compared with optimal a. Such a procedure remains valid and leads to a definition of pragmatic information for the case when the *a priori* probability distribution (whatever its nature, subjective or objective) is given. If it is not given, maximization of expected utility (MEU) must be replaced by some other criterion of decision-making which would then provide a standard of comparison for the elements of D.

Before we discuss this solution, let us summarize the four possible cases by means of the following table.

		A priori probabilities	
		not given	given
Information	not given	$u(a, s)$.	$E_s u(a, s)$
	given	$E_x u(d(x), s)$	$E_x E_s u(d(x), s)$

The expressions in the table are those to be maximized: in the first row, by the choice of a; in the second row, by the choice of d, (the letter E stands, of course, for the operator of taking expected value). The expression in the second column have no free variables except those subject to choice. In the expression in the first column, the free variable s appears, hence the necessity of a criterion. The discussion below is in terms of d, this case being the broader one in view of: $A \subset D$.

Generally speaking, a criterion K of decision-making associates with each d the number u_K(d), i.e. the K-value in utility of this decision function. Analogously, we shall have u_K(a) for each a. Thus, for instance, $u_K(d)$ may be the arithmetic mean of utilities $u\ (d, s)$ (where $K =$ the Laplace criterion) or their minimum ($K =$ maximum) or, for that matter, their expected value ($K =$ MEU, relative to subjective estimate $p(s)$ of chances), etc. For all such cases, pragmatic information is now defined[5] as

$$C(X, S;\ U, K) = \max_d u_K(d) - \max_a u_K(a)\,. \tag{32}$$

The concept is doubly relative: the decision problem U and to the criterion K of decision-making. The last relativization is a consequence of the fact *a priori* probabilities are no more a necessary part of the conceptual setup. They may, however, be additionally assumed, in which case they constitute, through MEU, one specialization of K.

Thus, for instance, if K = maximin (denoted by W) we have

$$C(X, S; U, W) = \max_{d} \min_{s} \sum_{x} u(d(x), s)p(x \mid s) - \max_{a} \min_{s} u(a, s). \tag{33}$$

And similarly for other criteria of decision-making.[6]

Let us remark in passing that the concept of pragmatic information may be used in assessing the criteria themselves. For it seems sensible to adopt the following postulate, to be satisfied by any action K of decision-making. If U is non-trivial (i.e., in A there is no action dominating all other actions in respect of utility) then $C(X, S; U, K) > 0$. In the light of the above, the maximum postulate, for instance, does not qualify as a reasonable principle: it can be shown[7] that pragmatic information relative to maximum is equal to zero only if U has a saddle point.

5 SOME SPECIAL CASES

Pragmatic information, as defined above, seems to exclude certain intuitively natural cases which do not involve probability and possibly do also without the variable X. For instance, the decision maker may be told simply which one is the actual state of the world or to which subset of S it belongs (the list of subset being specified in advance). Such information may be assumed to be true or not.

It is not difficult to see that such cases are covered by the concept introduced above. We obtain desired results by suitably interpreting the set X and imposing strong conditions on the probability distribution $p(x \mid s)$. A few examples will make this assertion clear. Let us interpret X as a partition of S, i.e. for each $x, x' \in X$:

$$\varnothing = x \subset S,\ U_x = S,\ x \neq x \Rightarrow x \cap = \varnothing. \tag{34}$$

If, in addition, the distributions $p(x \mid s)$ satisfy the condition:

$$\text{if } s \in x \text{ then for all } x' \neq x\text{: } p(x \mid s) > p(x' \mid s), \tag{35}$$

then it is natural to attach a semantic interpretation to the elements of X. Each x can be thought of as the statement that the actual state of the world belongs to the subset x of S. Condition (34) can be strengthened by the further assumption that each x contains exactly one element of S. This establishes a one-one correspondence between the two sets, hence they may be identified:

$$X = S. \tag{36}$$

again, (35) can be strengthened to:

$$\text{if } s \in x \text{ then } p(x \mid s) = 1 \tag{37}$$

Condition (37) ensures the truth of the statement that the actual state of the world belongs to x. Of course, the best possible situation is defined by (36) and (37).[8] In such a case, we obtain true and precise information concerning the state of the world. Information of this kind deserves the name of perfect: indeed, conditions (36) and (37) entail (26), which served in defining the notion.

Preceding remarks indicate the possibility of typology, according to the nature of the set X and the properties of the probabilistic relation between the elements of X and S. Obviously, if additional assumptions of the type mentioned above are made, all concretizations of the general expression (32) for pragmatic information simplify considerably. The case of perfect information has been investigated in more detail in [6]. Somewhat broader class, defined by (34) and (37), was considered in [8]. In both papers conditions were sought under which the pragmatic information (relative to different criteria) takes on extreme values. They are, of course, much simpler than those for the more complex case of $C(X, S; U)$.

NOTES

* Reprinted from *Theory and Decision* 1 (1974), D. Reidel Publishing Company, Dordrecht-Holland.

[1] Cf., for instance, Cherry [1]. This shows, incidentally, that the notion was loosely discussed at least some 15 years ago.

[2] Strictly speaking, d_{opt} is a representative of a (non-empty) class, as there may be more than one optimal decision function. However, if this is the case, they all satisfy (4), hence lead to the same utility (5).

[3] The same remark applies to the non-uniqueness of a_{opt} as to that of d_{opt}.

[4] See [2], cf. also [5].

[5] The definition can be justified in more detail in terms of the (non-negative) cost of obtaining information. Pragmatic information is then thought of as the highest cost, compatible with the optimality of a decision function belonging to the set D-A.

[6] Some of them are discussed in [4].

[7] For the proof (in the special case of perfect information), see [6].

[8] Strictly speaking, in (37) the element sign ought then to be replaced by identity.

REFERENCES

[1] C. Cherry, *On Human Communication: A Review, a Survey and a Criticism*, New York, 1957.

[2] C. G. Hempel, 'Inductive Inconsistencies', *Synthese* 12 (1960) 439–469

[3] J. Hintikka, 'The Varieties of Information and Scientific Explanation', in: *Logic, Methodology and Philosophy of Science III* (ed. by B. van Rotselaar and J. F. Staal), Amsterdam 1968.

[4] K. Szaniawski, 'Some Remarks Concerning the Criterion of Rational Decision-Making', *Studia Logica* 9 (1960) 221–239 [see this volume, pp. 114–127.].

[5] K. Szaniawski, 'A Pragmatic Justification of Rules of Statistical Inference', in: *The Foundation of Statement and Decisions* (ed. by K. Ajdukiewicz), Warszawa 1965 [see this volume, pp. 87–95.].

[6] K. Szaniawski, 'The Value of Perfect Information,' *Synthese* 17 (1967) 408–424 [see this volume, pp. 140–153.].

[7] K. Szaniawski, 'Information and Decision' (in Polish), *Zagadnienia Naukoznawstwa* 13 (1968) 69–79.

[8] K. Szaniawski, 'Pragmatic Value of Information' (in Polish), in: *Problemy psychologii matematycznej* (ed. by J. Kozielecki), Warszawa 1971.

19

TYPES OF INFORMATION AND THEIR ROLE IN THE METHODOLOGY OF SCIENCE

0 INTRODUCTION

Information is often said to be one of the aims of science. To put it in a somewhat metaphoric form: science is an information seeking process. If this is true, then the concept of information becomes of crucial importance for the methodology of science, because the use of a method in science would then have to be justified in terms of its efficiency in obtaining information. Attempts to do so are known in the literature. I would like to mention in this connection the well known book by I. Levi [2] and J. Hintikka's study [1]. In the present paper, certain results obtained so far will be systematized by means of the concept of pragmatic information (cf. Szaniawski's [5] and [6]; see also Nauta's [3]).

Since the main dividing line will be drawn between *apragmatic* and *pragmatic* varieties of information, I shall first restate briefly the principles on which they are based, then go on to discuss their mutual relations in the analysis of scientific procedures.

I APRAGMATIC INFORMATION

As the terminology suggests, apragmatic information is independent of relations that might exist between the language and its user. It is defined exclusively in terms of probabilities. In its basic form, *apragmatic information* is simply a decreasing function of probability. Out of the variety of such functions, two have been singled out because of certain desirable properties they possess (see [1] for details):

$$inf(t) = -\log p(t) \tag{1}$$

$$cont(t) = 1 - p(t), \tag{2}$$

where t is any statement.

A number of derived concepts can be defined, in terms of either *inf* or *cont*. Probably the most important is what Hintikka calls *transmitted information*, i.e. 'the amount of information that x conveys (contains) concerning the subject matter of *s*'. If will be denoted by *transinf*(x, s):

$$transinf(x, x) = inf(s) - inf(s \mid x). \tag{3}$$

In the above expression, $inf(s \mid x)$ is obtained from (1) by replacing the absolute probability of *s* by the conditional probability $p(s \mid x)$.

Let *s* and *x* in (3) range over the sets *X* and *S*, respectively. We assume the that $S(X)$ is such that the exclusive disjunction of all *s* in *S* (all *x* in *X*) is true. Under these conditions, *transinf* can be averaged. If we denote the resulting expression by $I(X, S)$ we have

$$I(X, S) = E_{p(s)}\, inf(s) - E_{p(x)}\, E_{p(s \mid x)}\, inf(s \mid x), \tag{4}$$

where E_p is, of course, the operator of the expected value with respect to the measure *p*. $I(X, S)$ is the famous *Shannon expression*, i.e. the difference between absolute entropy of *S* and the average conditional entropy of *S*, given *x* in *X*. It represents the average decrease of indeterminacy of *S*, due to the knowledge of *x*; in other words, the average amount of transmitted information (from *X* to *S* or vice versa: the expression is symmetric). An expression analogous to *I* can be obtained if *inf* is replaced by *cont*.

II PRAGMATIC INFORMATION

Numerical value of the expression $I(X, S)$ is entirely determined by the joint probability distribution function $p(x, s)$ where the sentential variables *x*, *s* range over the sets *X*, *S*, respectively. All the absolute and conditional probability distributions in (4) follow from $p(x, s)$.

Let us remark in passing that an important special case is obtained if $p(x, s)$ satisfies the condition

$$\bigwedge_x \bigvee_s [p(s \mid x) = 1]. \tag{5}$$

Then each *x* determines uniquely the answer to the question '*which s in S*?'. This fact justifies calling such (transmitted) information *perfect*.

Now, in order to be able to speak of pragmatic information, we must additionally introduce something else: we have to suppose that the consequences of a person's decision depend upon the question, which *s* in *S* is true. Let, therefore, *A* be a (finite) set of behavioral alternatives ('actions', 'strategies') the subject is about to choose from. And let *u* be a real function defined on $A \times S$, representing the subject's evaluation (his 'utility') of all possible pairs (a, s). The triple $U = \langle A, S, u \rangle$ connotes what is usually called *decision problem* in its normal form.

By *pragmatic information of X on S, relative to the decision problem U*, we shall mean the (average) improvement of decision in *U*, due to *X*. The definition is straightforward. If the decision-maker does not know which *x* in *X* is true, he simply maximizes *u*, averaged by means of $p(s)$. If he does know the true *x*, he maximizes *u*, relative to the conditional measure $p(s \mid x)$; and what he obtains in this way must be averaged over all x in X. It follows that the definition of pragmatic information, denote it by C, is

$$C(X, S \mid A, u) = E_{p(x)} \underset{a}{\text{Max}}\, E_{p(s\mid x)} u(a, s) - \underset{a}{\text{Max}}\, E_{p(s)}\, u(a, s). \tag{6}$$

By introducing a convenient shorthand

$$v(a, x) = E_{p(s\mid x)}\, u(a, s) \tag{7}$$

we can rewrite (6), more symmetrically, as

$$C(X, S \mid A, u) = E_{p(x)} \underset{a}{\text{Max}}\, v(a, x) - \underset{a}{\text{Max}}\, E_{p(x)}\, v(a, x). \tag{8}$$

For some purposes, still another representation will be more convenient. If we assume, as we did, that the sets X, S, and A are finite, then both $u(a, s)$ and the joint probability distribution function $p(x, s)$ are really matrices. Let us denote their product by w (in order to be able to perform the multiplication, we must first transpose the probability matrix):

$$[w(a, x)] = [u(a, s)] \cdot tr[p(x, s)], \tag{9}$$

then pragmatic information is easily seen to be

$$C(X, S \mid A, u) = \sum_x \underset{a}{\text{Max}}\, w(a, x) - \underset{a}{\text{Max}} \sum_x w(a, x). \tag{10}$$

Equation (9) and (10) suggest an easy way to study the relation between pragmatic information C and the Shannon expression I. In order to be able to compare the two, we must somehow make the utility function vanish, since it does not enter the definition of $I(X, 'S)$: this last expression was defined entirely in terms of $p(x, s)$. But if wc assume the utility matrix to be identity matrix, then w becomes equal to the transported p and we have C depending entirely on the joint probability distribution.

The meaning of the assumption just made is that the utility matrix is square and has 1 on the main diagonal, 0 elsewhere. This establishes the one-one correspondence between the sets A and S, hence A may be identified with S and we have the utility function (call it u_1) defined as follows

$$u_1(s, t) = \begin{cases} 1 & \text{if } s = t \\ 0 & \text{if } s \neq t \end{cases}. \tag{11}$$

A natural interpretation of such a decision problem is that, for any s and t in S, 'action' s consists in assuming s to be true, hence $u_1(s, t) = 1$ just in case the decision-maker is right, while $u_1(s, t) = 0$ in case he is wrong. The function u_1 is thus seen to be a special case of what we will call epistemic utility, following an already well established usage. In the next chapters we will discuss this concept more symmetrically.

For the moment, let me mark that $C(X, S \mid S, u_1)$ exhibits a strong similarity to $I(X, S)$; see [5] for details.

III EPISTEMIC DECISION PROBLEM

By suitably interpreting the set A and valuation u in U, we obtain the model of situation often found in science: the decision to be made consists in choosing a statement for acceptance. If the basic set of statements is S then the elements of A are truth-functional compounds, most conveniently: disjunctions, of some s in S (here and below I am following closely Levi's [2]).

If we identify, for simplicity of notation, a disjunction with the set of its elements, we may say that in an epistemic decision problem A is a family of subsets of S:

$$\varnothing \neq A \subseteq 2^s, \tag{12}$$

satisfying certain conditions. It is not, however, postulated that A be deductively closed. This does not mean that we reject the postulate of deductive closure of the rational subject's body of beliefs: having accepted an a in A, he is bound by this postulate to accept any a' in A such that $a \subset a'$. But insofar as his decision problem goes, a is the strongest statement that he is prepared to accept. To put it shortly, adopting a means: (i) the disjunction of all s in a is accepted as true; (ii) the same goes for all the supersets of a; (iii) if $a' \subseteq a$ then the disjunction of all s in a' is not accepted as true.

So much for the interpretation of elements of A. As to conditions which A is assumed to satisfy they are fairly obvious:

$$\text{If } a \in A \text{ then } a \neq \varnothing, \tag{13.1}$$

$$\bigvee_{a,\, a' \in A} a \neq a', \tag{13.2}$$

$$\bigcup_{a \in A} a = S. \tag{13.3}$$

In other words, a statement is accepted, the subject has a choice, and at least one a in A is true, whatever the actual s.

The exact character of A is, in each case, determined by the subject. Thus, he may want to know whether a particular s is true or not (verification of a simple hypothesis); in which case his A consists of two elements: $\{s\}$ and S-$\{s\}$. There are, of course, other possibilities in what is going to follow we shall limit our attention to two extreme cases:

$$A_1 = 2^s - \{\varnothing\} \tag{14.1}$$

$$A_2 = \{s\}_{s \in S} \tag{14.2}$$

The second case was already mentioned above: it was presupposed in (11). The first case allows any disjunction to be accepted as strongest, including S itself, the truth of which has been initially assumed. This is the case discussed in Levi's book (he interprets acceptance of S as the suspension of judgment). If A is defined by (14.1) it will be called *unrestricted*; and *restricted* if it is defined by (14.2).

Given the above interpretation of A, what shape is the function u to have in an epistemic decision problem? The answer depends on the goal the subject wants to achieve. Insofar as the epistemic decision problem is a model of scientific procedure, the utility function u may be said to define the aims of science (thus Levi considers his book 'an essay on induction and the aims of science').

We are here interested in science as purely cognitive activity, thereby excluding all cases of applied research from the scope of our inquiry. Even so, the characterization of the aims of scientific research remains a fairly complex task, because: (a) there is a variety of possible aims, depending on the context; (b) in some cases, the research is supposed to achieve two conflicting aims simultaneously; (c) quite often it is not possible to go beyond a verbal and vague description of what the scientist wants to achieve.

I will try below to discuss the epistemic utility in more detail. It seems, however, that there is one postulate any epistemic utility function ought to satisfy. Namely, that every error be valued less than any non-error. Obviously, for a given s, the adoption of a is an error iff $s \notin a$. Hence, the postulate on u has the following form:

$$\bigwedge_{s} \bigwedge_{a, a'} [\text{if } s \in a \text{ and } s \notin a' \text{ then } u(a, s) > u(a', s)]. \tag{15}$$

One simple way to ensure that (15) is satisfied can be described as follows. Suppose that the elements of A are evaluated by means of a non-negative function h, irrespective of whether they are erroneous or not. Then the following definition of u in terms of h will make u satisfy (15):

$$u(a, s) = \begin{cases} h(a) & \text{if } s \in a \\ h(a) - \operatorname*{Max}_{a} h(a) & \text{if } s \notin a \end{cases} \tag{17}$$

We then have $h(a) > 0 > h(a) - \text{Max}_a h(a)$, with the possible exception of the boundary case. The postulate (15) is a formal expression of the fact that truth takes precedence over any other epistemic value. We now come to the discussion of those 'other values'.

IV GLOBAL THEORIZING

The distinction between global and local theorizing is due to Hintikka [1]. In his own formulation: 'one of the most important distinctions here is between, on one hand, a case in which we are predominantly interested in a particular body of observations e which we want to explain by means of a suitable hypothesis h, and, on the other hand, a case in which we have no particular interest in our evidence e but rather want to use it as a stepping stone to some general theory h which is designed to apply to other matters, too, besides e. We might label these two situations as case of *local* and of *global theorizing*, respectively. Often the difference in question can also be characterized as a difference between explanation and generalization, respectively'. (p. 321)

We are now going to try to formalize the case of global theorizing, in the sense of defining a utility function which would express this particular goal of research. It seems, however, that global theorizing has at least two conflicting goals. One is *truth*, the other is *information*.

Suppose first that truth is the only aim. This means that all cases of accepting a true statement are treated alike, similarly all cases of committing an error, and the first are preferred to the second, in accordance with (15). The utility function is, therefore, a generalization of (11):

$$u_1(a, s) = \begin{cases} 1 & \text{if } s = a \\ 0 & \text{if } s \neq a \end{cases} \tag{17}$$

If A is restricted in the sense of (14.2), pragmatic information of X on S, relative to u_1, is:

$$C(X, S \mid A_2, u_1) = E_{p(x)} \operatorname*{Max}_{s} p(s \mid x) - \operatorname*{Max}_{2} p(s). \tag{18}$$

As indicated above, the expression behaves in a very similar way to $I(X, S)$. On the other hand, if A is unrestricted, i.e. defined by (14.1), C becomes equal to 0:

$$C(X, S \mid A_1, u_1) = 0. \tag{19}$$

The explanation is fairly simple. If truth is the only aim, the sense of (17) then the maximal disjunction of all elements of S dominates any other a in A_1, i.e. is rated higher by u_1, under any s. Acceptance of a tautology (equivalently, suspension of judgment) is the best in an absolute sense, hence any information becomes superfluous.

Another obvious aim of global theorizing is information. For reasons which have been sketched by Hintikka, *cont* is in this context a more appropriate measure than *inf*. Suppose, therefore, that elements of A are appraised in terms of *cont*. If the new utility function, say u_2, is to satisfy the postulate (15) then the function $h(a)$ in (16) will simply become cont(a) and we have the following definition of u_2:

$$u_2(a, s) = \begin{cases} cont(a) & \text{if } s \in a \\ cont(a) - 1 & \text{if } s \notin a \end{cases} \tag{20}$$

which gives the following expression for the information of X on S, relative to u_2:

$$C(X, S \mid A, u_2) = E_{p(x)} \operatorname*{Max}_{a}[p(a \mid x) - p(a)]. \tag{21}$$

In the above formula, $p(a \mid x)$ is of course, a shorthand:

$$p(a \mid x) = \sum_{s \in a} p(s \mid x), \tag{22}$$

and similarly for $p(a)$.

The expression in the square brackets, to be maximized by a suitable choice of a, has been extensively discussed in the literature (cf. Hintikka [1], p. 325–329). Interpreted as the increase in the probability of a, due to the evidence x, it forms the basis of many measures of evidential support. Formula (21) shows that its maximal value, averaged over all x's and in X, is identical with pragmatic information of X on S, if global theorizing's aim is defined as maximization of *cont*, subject to the condition (15).

If the joint probability distribution satisfies (5), i.e. if the information of X on S, is perfect, the expression (21) depends on the marginal distribution of s:

$$C(X, S \mid A, u_2) = 1 - E_{p(s)} p(s), \tag{23}$$

which becomes maximal in the case of equidistribution:

$$\operatorname*{Max}_{p(s)} C(X, S \mid A, u_2) = 1 - \frac{1}{n}, \tag{24}$$

where n is the number of elements in S.

The two aims (search for truth, search for information) are in obvious conflict. From the first point of view, an element of A is the more highly valued, the more probable it is. Whereas *cont* is inversely related to probability. In actual practice some kind of compromise between these two conflicting goals is established. Its exact nature depends upon the subject's (the scientist's) decision. In particular, upon his willingness to risk (cf. Levi's expression: 'gambling with truth'). Let us, therefore, form a 'mixed' utility, say u_2, by means of Levi's parameter q, satisfying the inequality $0 \leq q \leq 1$.

$$u_3 = (1 - q) \cdot u_1 + q \cdot u_2 . \tag{25}$$

As a result we obtain

$$u_3(a, s) = \begin{cases} 1 - q \cdot p(a) & \text{if } s \in a \\ - q \cdot p(a) & \text{if } s \notin a \end{cases} \tag{26}$$

The interpretation of q is obvious: the higher its value, the more willing the scientist is to risk an error in order to win information in the sense of *cont*. The generalized expressions for pragmatic information are, in the case of unrestricted and restricted A, respectively, as follows.

$$C(X, S \mid A_1, u_3) = E_{p(x)} \operatorname*{Max}_{a} [p(a \mid x) - q \cdot p(a)] - (1 - q) \tag{27}$$

$$C(X, S \mid A_2, u_3) = E_{p(x)} \operatorname*{Max}_{s} [p(s \mid x) - q \cdot p(s)] - (1 - q) \cdot \operatorname*{Max}_{s} p(s). \tag{28}$$

If p satisfies (5), the above reduce to

$$C(X, S \mid A_1, u_3) = q \cdot [1 - E_{p(s)} p(s)] \tag{29}$$

$$C(X, S \mid A_2, u_3) = 1 - q \cdot E_{p(s)} p(s) - (1 - q) \cdot \operatorname*{Max}_{s} p(s). \tag{30}$$

It is easy to see that Levi's epistemic utility function (cf. [2], p. 81) is a special case of u_3, obtained by means of postulated equidistribution on S. If we denote such a measure on S (used by Levi only to define *cont*) by m, and the resulting utility function by u_4 we have

$$u_4(a, s) = \begin{cases} 1 - q \cdot m(a) & \text{if } s \in a \\ -q \cdot m(a) & \text{if } s \notin a \end{cases} \tag{31}$$

In the case of restricted A, $m(a)$ becomes constant in a, which makes u_4 identical with u_1, up to linear transformation. In other words, if the subject decides in advance to have one of the logically strongest statements accepted, he behaves, in terms of u_4, as if his only aim were truth (no matter what the value of q is). The genuine compromise between truth and *cont* (defined in terms of m) is arrived at if A is unrestricted. This is the case investigated by Levi. Pragmatic information is then equal to

$$C(X, S \mid A_1, u_4) = E_{p(x)} \operatorname*{Max}_a [p(a \mid x) - q \cdot m(a)] - \operatorname*{Max}_a [p(a) - q \cdot m(a)]. \tag{32}$$

To close this section, a remark concerning the epistemic utility u_2, which played an essential role in the above analysis. Let us compare it with the utility defined by Hintikka in an analogous context ([1], p. 325).

Hintikka describes the aim of global theorizing as follows (I am changing the symbols in an inessential way): 'If a is true, we gain the (substantive) information *cont*(a) by adopting a. If a is false, we lose the information which we could have gained by opting for $\sim a$, rather than a. According to what was just said, our net utility in this case is, therefore, -*cont*($\sim a$)'. Now, it is easy to see that: -*cont*($\sim a$) = *cont*(a) $- 1$. It thus turns out that Hintikka's utility is identical with u_2, which has been obtained above in another way.

V LOCAL THEORIZING

The starting point is now different. We assume a statement, call it e, as given and the problem is, how to explain e by suitable choice of an s in S (the discussion will be limited to restricted A). Following Hintikka, we associate the meaning of the word 'explain' with the notion of transmitted information: e is the better explained by s, the more information s conveys concerning the subject matter of e.

Thus, *transinf*(s, e) is a means of defining the purpose of *local* theorizing. It, seems, however, that simply to identify this purpose with maximization of *transinf* would not be an adequate description of what actually takes place. From the scientist's point of view it is not immaterial whether the intended explanation is true or not. If, therefore, the utility function u_e is to define the purpose: to explain e, it ought to satisfy the postulate (15). In order to assure this, we apply (16) once more,[1] this time making h equal to *transinf*. In view of

$$\operatorname*{Max}_s \mathit{transinf}(s, e) = \operatorname*{Max}_s \log \frac{p(e \mid s)}{p(e)} = -\log p(e), \tag{33}$$

we have

$$u_e(s, t) = \begin{cases} \log p(e \mid s)/p(e) & \text{if } s = t \\ \log p(e \mid s) & \text{if } s \neq t \end{cases}. \tag{34}$$

Averaged over all t in S, this gives

$$E_{p(t \mid e)} u_e(s, t) = \log p(e \mid s) - p(s \mid e) \log p(e). \tag{35}$$

If $p(s)$, i.e. the prior distribution of s, is uniform then the maximization of the above expression is equivalent to maximization of the likelihood of e, i.e. of $p(e \mid s)$. In other words, the maximum likelihood principle guides the choice of explanation if all explanations are *a priori* considered equiprobable.[2] Obviously, another interpretation of local theorizing is also possible.

It may be thought that the process of explanation is first carried out without considering the logical value of the explanans. Maximization of *transinf* takes place and after it has led to the selection of an s, an independent test of its truth is made. According to this view, maximum likelihood principle leads to the selection of the explanans, because it is equivalent to the maximization of transmitted information (cf. Hintikka, [1], p. 322). Then, however, maximum likelihood principle could not be called a rule of acceptance, contrary to its current interpretation, unless we are prepared to accept statements regardless to their logical value.

VI CONCLUDING REMARKS

On the preceding pages I tried to show how the various goals of scientific research can be defined in terms of information, in its many variants. In the case of global theorizing, the (average) value of the data, i.e. of the elements of set X, can be calculated, with respect to such a goal. This is what I proposed to call *pragmatic information*, relative to different kinds of epistemic decision problem.

It was presupposed throughout that all the probabilities are given, i.e. the joint probability distribution function of the relevant variables is known. This assumption has been, and still is, the subject of controversy. It has consistently been maintained by some writers that the only probabilities actually given are the conditional ones of the type $p(x \mid s)$. In other words, that we ought to dispense with the marginal distribution $p(s)$. What would be the consequences of this view?

Obviously, certain concepts would have to go, viz. those which are defined in terms of $p(s)$. It follows that the goals of research would have to be redefined. The concept of pragmatic information could be retained in a different form, relativized to a criterion of decision making, such as *maximin, minimax loss, etc.* The way to do this is sketched in [5].

NOTES

1 We are here making the rather natural assumption that only such statements are possible candidates for explaining e which are positively associated with it. It follows that $transinf(s, e)$ is positive for any s in S.

2 The connection between the maximum likelihood principle and the assumption of equiprobability of hypotheses has been investigated in [4].

REFERENCES

[1] J. Hintikka, 'The Varieties of Information and Scientific Explanation', in: *Logic, Methodology and Philosophy of Sciences* III, van Rootselaar and Staal, (ed.) North Holland, 1968.

[2] I. Levi, *Gambling with Truth. An Essay on Introduction to the Aims of Science*, Alfred A. Knopf, 1967.

[3] D. Nauta, Jr, *The Meaning of Information*, Mouton, 1972.

[4] K. Szaniawski, 'The Maximum Likelihood Principle: An Evaluation'. (in Polish) in: *Rozprawy Logiczne. Essays in Honour of K. Ajdukiewicz*, PWN, Warsaw, 1964.

[5] K. Szaniawski, 'Two Concepts of Information', *Theory and Decision*, 5, 1974. [see this volume, pp. 157–166.]

[6] K. Szaniawski, 'Information and Decision as Tools of Philosophy of Science', *Danish Yearbook of Philosophy*, Vol. 10, 1973. [see this volume, pp. 8–18.]

20
INFORMATION IN DECISION-MAKING SOME LOGICAL ASPECTS

1 INTRODUCTION

In this paper I shall try to describe formally, in the simplest possible way, the role information plays in decision-making. Intuitively, this is rather well known from everyday experience. We all look for information when we face a decision problem, particulary if the decision is both important and difficult. In most cases, we know how to use that information and how to appraise its value in terms of the improvement it brings into our decision making. The definitions below seem, therefore, to be based on firm ground. Nevertheless, a few words of explanation might help to avoid misunderstanding as to the purpose of the present paper. Information, if the word is taken in a broad sense, enters the process of decision making in more than one way. Some of them lend themselves to formal treatment, whereas others have not been formalized so far and it is even doubtful that they ever will.

Thus, for instance, information may result in transforming an ill-defined decision problem into a well-defined one. This happens when the decision maker learns that he has an option he had not previously considered; or that, on the contrary, he cannot perform an action he previously thought possible. Other components of the decision problem are also subject to corrections, in the light of information obtained by the decision-maker.

Important as it undoubtedly is, information of that kind can hardly be formalized. Its definition would have to describe properties of something that is either not known or, at best, known very vaguely. For this reason, all non-verbal definitions of information start from a clearly defined (closed) set of alternatives.

My second remark concerns the character of actions in the description of decision problem. Here, they will be treated as strategies, i.e. as points in the space of alternatives the decision-maker chooses from. The same applies to the values of uncontrolled variable, whether they are 'states of the world' or the acts of intelligent opponent. In other words, the decision problem will have 'normal form'. Such an approach has non-negligible consequences for the problem of information. Had the decision problem been given extensively, e.g.

in the form of a game tree, information might consist in the reduction of certain fragments of that tree. By treating a strategy as an indivisible whole, we rule out this possibility. To be sure, this is less realistic, and it makes the theory more difficult to apply. On the other hand, for extensive problems defined quite generally, the conceptual structure becomes very complex indeed.

The adjective 'logical' in the tiltle of the paper indicates that emphasis will be laid on conceptual structure and justification of criteria, rather than on the more technical problem of how to find the optimal solution in specific conditions. Thus, for instance, all the sets spoken about are assumed to be finite.

The central concept of this paper is that of pragmatic value of information.[1] In spite of being simple almost to the point of triviality, this concept has a number of applications which justify its analysis. Here are a few examples.

Almost all decision problems generate secondary questions which have something to do with information. One of them is, whether to invest in the search for information or not. If the answer is 'invest' then: to what extent? And what kind of information to look for? And how to make the best use of it? An attempt to make these questions more precise leads to the concept of pragmatic value of information.

Also, certain procedures in science can be explained in terms of pragmatic information if utilitity (or loss) function is so chosen as to represent the scientist's objective. Of course, this applies to the problems of choice only, not to those of discovery. The former, however, are not infrequent in science: inductive interference (thus, mathematical statistics) belongs to procedures of this type. Another example is the choice of certain indicators, such as measures of association, if they are thought of as tools for a given purpose, viz. prediction. The concept of pragmatic value of information seems to be useful in this context.

The above is confirmed by the growing number of publications on this subject. See for instance, Marschak [2], McGuire [3], Ponssard [4], Ponssard and Zamir [5].

The exposition below is based on previous publications by the present author: Szaniawski [6–11]. The arrangement of the material is as follows. The concept of the pragmatic value of information is based on that of question and of decision problem. Following the distinction between categorical and probabilistic questions, the discussion of information's pragmatic value is first done without probability, then using probabilistic data. Brief mention is made about the relation between pragmatic information and the classical, entropy-based concept. Finally, the possibility of extending the analysis to games of strategy is outlined.

2 QUESTIONS

For reasons that have been mentioned above, any formal theory of information presupposes a well-defined set of alternatives, pairwise exclusive and jointly exhaustive. We will denote it by X. Its elements may be thought of, in the simplest case, as the values a given variable can assume. More generally, they are alternative states of the world, of any degree of complexity.

The most obvious specimen of a question concerning X is the straightforward inquiry: which x? The set of answers to this questions is identical with the set X itself. But we also learn something about actual state of the world if we get to know whether it belongs to some proper subset of X, say X', or not. Here the set of answers consist of two elements: the set of X' and its complement. Still weaker knowledge is represented by having the answer to such a question which directly concerns a set of 'symptoms', probabilistically related to elements of X. This happens to be the most general case. It will, therefore, provide the basis for the definition below.

Let Y be a set of alternatives, pairwise exclusive and jointly exhaustive. Let $p(y \mid x)$ be a probability distribution function on Y, parametric (or conditional) with respect to x; $x \in X$. By *a question on* X we shall mean the order pair

$$\langle Y, \{p(y \mid x)I\}_{x \in X}\rangle \tag{1}$$

The second element of this pair is, of course, the family of all distributions $p(y \mid x)$, for $x \in X$.

It is additionally assumed that if the marginal probability distribution $p(x)$ on the set X is given then the probabilistic relation between X and Y may be represented by the joint probability distribution function

$$p(x, y) = p(y \mid x)p(x) \tag{2}$$

The definition (1) is purposely weak: it does not presuppose that the so-called *a priori* probabilities of the type $p(x)$ are known. It is also possible to add such an assumption. Whether to do it or not, is a much debated issue which we do not intend to discuss here. Non-probabilistic questions are easily obtained by imposing suitable conditions on (1). A question is *categorical* if it satisfies

$$\bigwedge_z \bigvee_y (p(y \mid x) = 1). \tag{3}$$

Such zero-one probability distributions generate a partition of the set X. To any of $y \in Y$ we associate the set X_y of those states under which the probability of y is 1:

$$X_y = \{x\colon p(y \mid x) = 1\}. \tag{4}$$

If y actually obtains, then the true state of the world belongs to X_y. The sets X_y form a partition of X, since they are pairwise disjoint and sum up to X. A categorical question thus amounts to asking which subset X_y of X the actual state belongs to.

A still more specific case is obtained if we add to (3) the condition that for any y there be exactly one such x which makes y a sure event:

$$\bigwedge_y \bigvee_x p(y \mid x) = 1. \tag{5}$$

Then, of course, a one-one correspondence between the sets X and Y is established and any y specifies uniquely the actual state of the world. As this obviously constitutes the best possible case, a question satisfying conditions (3) and (5) will be called *perfect*.

If prior probabilities $p(x)$ are assumed, and we have, through Bayes's theorem, the posterior probabilities $p(x \mid y)$, then the condition defining perfect question becomes

$$\wedge_y \vee_x (p(y \mid x) = 1), \tag{6}$$

which is, of course, simpler than the conjunction of (3) and (5).

Probabilistic questions are more general than non-probabilistic ones (i.e. categorical), since the letter can be obtained as a special case of the former by means of (3).

As indicated above, information's pragmatic value will be understood here as *the value, relative to a decision problem, of having a question answered.* The type of the question is, of course, not indifferent from this point of view. We will begin our discussion the simpler, i.e. non-probabilistic questions. Before we proceed to do that, a few words on the concept of decision problem are necessary. The description will be done in telegraphic style, since it refers to generally known notions.

3 DECISION PROBLEM

The paradigm-case amounts essentially to this: in a well-defined decision problem the subject knows what *can* happen as the result of his decision; he does not know what actually *will* happen. Everything else is assumed to be explicitly given. Hence, a decision problem L is defined as

$$L = \langle A, X, l \rangle, \tag{7}$$

where: A is the decision-maker's choice set (the set of his possible acts); X is the set of possible states of the world influencing the outcome of his decision; l is a real function on $A \times X$, interpreted as the loss incurred when the decision-maker chooses an a out of A and the state of the world is an x in X.

The subject's purpose is, of course, to minimize l.

Out of several criteria that have been proposed for doing this, let us recall a few.

$$\text{M(inimization) of E(xpected) L(oss): Minimize } E_{p(x)}\, l(a, x). \tag{8}$$

$$\text{Minimax: Minimize } \operatorname*{Max}_x l(a, x). \tag{9}$$

$$\text{Minimax regret: Minimize } \operatorname*{Max}_x w(a, w), \tag{10}$$

where $w(a, x) = l(a, x) - \operatorname*{Min}_a l(a, x)$.

$$\text{The Laplace criterion: Minimize } \sum_x l(a, x)/n, \tag{11}$$

where n = the number of elements in X.

a-criterion of Hurwicz: Minimize $aM(a) + (1 - a)m(a)$, (12)

where $M(a) = \text{Max}_x\, l(a, x)$, $m(a) = \text{Min}_x\, l(a, x)$, $0 \leq a \leq 1$.

We notice that MEL is the only one that uses prior probabilities. The Laplace criterion is the meeting ground of the two approaches: it can be interpreted either as a special case of MEL, or as a criterion denying the use of prior probabilities. The Hurwicz criterion is a parametric one: once the decision maker has settled on some value of a, representing his attitude towards uncertainty (in this case, his 'degree of pessimism'), the expression in (12) becomes uniquely determined.

4 INFORMATION GENERATED BY CATEGORICAL QUESTIONS

We will introduce the basic notions by means of a simple numerical example. Let the decision problem be

	x_1	x_2	x_3	x_4
a_1	4	3	1	2
a_2	0	2	5	7

Let $P = \{x_1, X_2\}$ be the following partition of $X: X_1 = \{x_1, x_2\}, X_2 = \{x_2, x_4\}$. The categorical question corresponding to this partition is: which of the two subsets, X_1 or X_2, does the actual state belong to? The problem now is to determine the value, for the decision-maker, of having this question answered. In other words, how much does he stand to gain when he is in the position to make his decision depend on the answer to the question above?

He can do it in four ways, since there are four function from P to A. Obviously, one such function represents the optimal use of information, therefore it should be adopted by the decision maker. It consists in choosing a_2 if the answer is X_1, choosing a_1 if the answer is X_2. The result, tabulated against the original problem, is as follows.

	x_1	x_2	x_3	x_4
a_1	4	3	1	2
a_2	0	2	5	7
d_0	0	2	1	2

We have denoted by d_0 the optimal decision function; more generally, a *decision function associates* some action in A with each answer to the question.

It is clear from the the above that information generated by a categorical question makes it possible to 'break up' the original actions into segments determined by the question and to put the segments together again in such a way that the resulting whole constitutes (in general) an improvement upon each element in A. How to assess this improvement? Obviously, d_0 ought to be compared with the best action in A. Any criterion of decision making provides the means of doing this.

Suppose, for instance, that the decision maker intends to apply MEL (he is a 'Bayesian') and his prior distribution over X is $p = \langle 0.2, 0.3, 0.1, 0.4\rangle$. The expected loss for a_1, a_2 is 2.6, 3.9, respectively; while that for d_0 is 1.5. We notice that a_1 is optimal in A. Relative to a_1, the decrease of expected loss, due to the use of d_0, is $2.6 - 1.5 = 1.1$ It seems natural to say that 1.1 is the *value of information* (*or pragmatic information*) generated by the question P, with respect to the above decision problem and relative to the prior distribution p; it is, of course, presupposed that the criterion applied is MEL.

For illustration's sake, let us repeat the argument in terms of some other criteria mentioned above. If minimax is applied we have $\text{Max}_x\ l(a_1, x) = 4$, $\text{Max}_x\ l(a_2, x) = 7$, therefore a_1 is better. On the other hand, $\text{Max}_x\ l(d_0, x) = 2$. It follows that pragmatic information, relative to minimax, is $4 - 2 = 2$.

In the case of minimax regret, we transform the matrix of losses into the matrix of regrets $w(a, x)$;

	x_1	x_2	x_3	x_4
a_1	4	1	0	0
a_2	0	0	4	5
d_0	0	0	0	0

We have $\text{Max}_x\ w(a_1, x) = 4$, $\text{Max}_x\ w(a_2, x) = 5$, hence pragmatic information generated by P is, in terms of minimax regret, $4 - 0 = 4$.

Generalizing the above argument, we obtain the following definitions. Let $L = \langle A, X, l\rangle$ be a decision problem and let $P = \{X_y\}_{y \in Y}$ be a partition of X, i.e. a family of non-empty, pairwise disjoint subsets, summing up to X. We will consider pragmatic information, generated by the question: Which of X_y does the actual state of the world belong to? The concept to be defined depends, of course on the state of criterion of decision making.

Let $p(x)$ be a probability distribution on X. We will denote by $C(X:P; L, p)$ the *pragmatic value of categorical information generated by P, relative to the decision problem L and the prior distribution* $p(x)$. We identify with the decrease of average loss, due to the optimal use, by the decision maker, of the answer to the question P. Hence the following definition:

$$C(X, P; L, p) = \underset{a}{\text{Min}} \sum_{x} l(a, x)p(x) - \sum_{y} \underset{a}{\text{Min}} \sum_{x \in Xy} l(a, x)p(x). \tag{13}$$

By means of the abbreviation

$$e(a, y) = \sum_{x \in Xy} l(a, x)p(x), \tag{14}$$

we obtain more symmetric form of (13):

$$C(X: P; L, p) = \underset{a}{\text{Min}} \sum_{y} e(a, y) - \sum_{y} \underset{a}{\text{Min}}\ e(a, y). \tag{15}$$

If the criterion employed is minimax, i.e. the decision functions are compared with the actions in A in terms of the maximum loss associated with them, then the same argument applies and it leads to the following expression for the value of information:

$$\begin{aligned} C(X: P; L, p, \text{minimax}) &= \operatorname*{Min}_a \operatorname*{Max}_x l(a, x) \\ &\quad - \operatorname*{Max}_y \operatorname*{Min}_a \operatorname*{Max}_{z \in X_y} l(a, x) \\ &= \operatorname*{Min}_a \operatorname*{Max}_y m(a, y) - \operatorname*{Max}_y \operatorname*{Min}_a m(a, y), \end{aligned} \tag{16}$$

where

$$m(a, y) = \operatorname*{Max}_{x \in X_y} l(a, x). \tag{17}$$

An analogous formula is obtained if the loss function l is replaced by 'regret function' w. And similarly for other criteria.

The limiting case of categorical information is, of course, generated by such a partition, here designated by Pt, which is equivalent to the set X itself, i.e. each element in Pt contains exactly one x from X. The question corresponding to Pt is simply 'which x?' and to have it answered represents the best type of information possible; we have called it 'perfect'.

For perfect information, definitions of the type given above simplify considerably. Here are some examples:[2]

$$\begin{aligned} C(X, Pt; L, p) &= \operatorname*{Min}_a E_{p(x)} l(a, x) - E_{p(x)} \operatorname*{Min}_a l(a, x) \\ &= \operatorname*{Min}_a E_{p(x)} w(a, x) \end{aligned} \tag{18}$$

Thus, the value of perfect information, relative to the prior distribution $p(x)$, turns out to be a minimum average regret. (The symbol E represents, of course, the operator of taking expected value by means of the probability measure indicated.)

$$C(X, Pt; L, \text{minimax}) = \operatorname*{Min}_a \operatorname*{Max}_x l(a, x) - \operatorname*{Max}_x \operatorname*{Min}_a l(a, x). \tag{19}$$

The above is, of course, the difference between so-called upper and lower value of the game (two-person, zero-sum), defined in terms of payoffs $l(a,x)$ to the second 'player', i.e. the one controlling x. The minimal value of (19), i.e. zero, obtains if it is such that there exists a saddle point; by saddle point we mean such a pair $\langle a_0, x_0 \rangle$ that $l(a_0, x_0)$ is maximal for a and minimal for x. Incidentally, it is easy to verify that the pragmatic value of information is non-negative for any criterion and any decision problem.

$$C(X, Pt; L, \text{minimax regret}) = \operatorname*{Min}_a \operatorname*{Max}_x w(a, x), \tag{20}$$

since the 'regret' w associated with optimal decision function based on perfect information, is uniformly zero.

As a sample of problems that can be asked concerning the value of perfect information, let us consider the following one. Given a criterion of decision making, say minimax, the value of perfect information is entirely determined by the structure of the decision problem. Now, suppose that the loss function has been normalized to the to zero-one interval. This makes it possible to speak of the maximal and minimal vlaue of perfect information, with respect to the decision problem. Minimum is, of course, zero; a sufficient and necessary condition for this case has already been given. What about the maximum?

It is easy to see that perfect information's value, in terms of minimax, is equal to 1 iff (i) zero loss is possible under any x, and (ii) for all a, the worst (i.e. the loss equal to 1) can happen. A good instance is represented by he decision problem defined as follows. The sets A and X are equipotent and there exists such a one-one correspondence $\leftrightarrow$ between their elements that

$$l(a, x) = \begin{cases} 0 & \text{if } a \leftrightarrow x. \\ 1 & \textit{otherwise} \end{cases} \tag{21}$$

Decision problem of this type has a natural interpretation. For any a and x, if $a \leftrightarrow x$ then a means: accepting x as true. Hence, (21) defines the simplest type of 'epistemic' loss function, with all possible errors causing the loss equal to 1. We will denote this decision problem by E.

Similarly for other criteria: E maximizes the value of perfect information. In the case of MEL, the value also depends on the prior distribution $p(x)$, and we have

$$C(X, Pt; E, p) = 1 - \operatorname*{Max}_{x} p(x). \tag{22}$$

We notice that perfect information reaches its maximal value, viz. $1 - 1/n$, if all the elements of X are, *a priori*, equally probable. On the other hand, $C = 0$ if some x has the prior probability 1. Concerning the last case, one may say that the decision maker has no use for information if he is subjectively certain about the state of the world.

5 INFORMATION GENERATED BY PROBABILISTIC QUESTIONS

It will be remembered that, generally speaking, the connection between the sets X and Y is of a probabilistic nature: it is represented by the probability distributions $p(y \mid x)$. If, in addition to that, an *a priori* distribution $p(x)$ on X is given then the pair $\langle X,Y \rangle$ can be treated as a two-dimensional random variable, with the joint probability distribution $p(x, y)$ fully determined by (2). This corresponds to the case when the criterion of solving the decision problem is MEL: the distribution $p(x)$ represents the decision-maker's estimate of the relative chances of elements in X.

We are going to limit our discussion to this case only (cf. Szaniawski [10]). A fuller treatment, in terms of other criteria, is both complex and unrewarding.

In analogy with the argument concerning the categorical case and leading to the formula (13), we compare the result of the optimal use of information with the outcome of the optimal action in A. When the decision is made without the benefit of information, the minimum average loss is

$$\operatorname*{Min}_{a} E_{p(x)}\, l(a, x). \tag{23}$$

Now, the optimal use of any y consisting in adopting such an a which minimizes the loss, averaged by means of the '*posteriori*' probability measure $p(x \mid y)$. This quantity is then averaged over all y in Y, which finally yields

$$E_{p(y)} \operatorname*{Min}_{a} E_{p(x \mid y)}\, l(a, x). \tag{24}$$

The difference between (23) and (24) is the pragmatic value of information when the criteria used is MEL:

$$C(X, Y; L) = \operatorname*{Min}_{a} E_{p(x)}\, l(a, x) - E_{p(y)} \operatorname*{Min}_{a} E_{p(x \mid y)}\, l(a, x). \tag{25}$$

if we adopt the shorthand

$$k(a, y) = E_{p(x \mid y)}\, l(a, x) \tag{26}$$

we simplify this expression to the formula:

$$C(X, Y; L) = \operatorname*{Min}_{a} E_{p(x)}\, k(a, y) - E_{p(y)} \operatorname*{Min}_{a} k(a, y). \tag{27}$$

The above expression reaches its minimal value, i.e. zero, iff there exists such an action a^* in A which is optimal *under all* y:

$$k(a^*, y) = E_{p(x \mid y)}\, l(a^*, x) \leq E_{(x \mid y)}\, l(a, x) \quad \text{for all } a \text{ and all } y. \tag{28}$$

The existence of such an action makes superfluous the information generated by Y. It is easily seen that if in the decision problem L there exists an action a^* dominating every other action, i.e. such that

$$l(a^*, x) \leq l(a, x) \quad \text{for all } a \text{ and all } x, \tag{29}$$

then this a^* satisfies also (28), hence reduces the pragmatic value of information to zero. Also, if Y is stochastically independent from X, i.e.

$$p(x, y) = p(x)p(y) \quad \text{for all } x \text{ and all } y, \tag{30}$$

then again (28) is satisfied by some a^*, viz. by the action optimal in terms of $p(x)$. Finally let us see what happens to $C(X : Y; L)$ if the epistemic problem E is substituted for L. We then obtain

$$C(X\colon Y; E) = \sum_{y} \operatorname*{Max}_{x} p(x, y) - \operatorname{Max} p(x). \tag{31}$$

This expression, defined entirely in terms of the probabilistic relation between X and Y, behaves in a closely similar way to the familiar information measure, defined by Shannon as a difference between the 'absolute' entropy of X and its average conditional entropy:

$$I(X: Y) = H(X) - E_{p(y)} H(X \mid y). \tag{32}$$

A more detailed account is found in Szaniawski [10].

6 THE ROLE OF INFORMATION IN SITUATIONS OF CONFLICT

It seems rather natural to inquire what kind of problems are generated when the concept of information's pragmatic value is carried over into the field of games. For this purpose, it is enough to interpret the elements of the set X as the strategies of an intelligent opponent whose valuations are known. Formally, let $G = \langle A, B, u_1, u_2 \rangle$ be a two persons game; $A(B)$ is the set of strategies and $u_1(u_2)$ is the utility function of the first (second) player, defined over $A \times B$. We are now speaking of utilities (i.e. negative losses), in accordance with language of the game theory. Therefore, minimization is replaced by maximization.

Perfect information only will be considered here. In other words, information about a player's choice will consist in pointing out an element in the set of his strategies. It is also assumed that the information is believed in. This assumption could, in some cases, be incompatible with the hypothesis of rationality of the players. In order to eliminate such a possibility we postulate that if e.g. player 1 is informed that 2 has chosen a strategy b then: (i) b is admissible, i.e. not dominated strongly by some other strategy in B; (ii) in case it is to the disadvantage of 2 to inform 1 of his choice of strategy, 1 has grounds to believe that he is obtaining the information without 2's knowledge. He can then act upon the information, without violating the principle of rationality of the players.

The seemingly innocuous extension of our conceptual apparatus to the case of two-person games creates, in fact, problems of considerable complexity. The present remarks cannot do more than indicate some of them.[3]

Probably the simplest case consists in player 1 obtaining the information about player 2's choice *without the latter's knowledge*. This information makes it possible for 1 to maximize u_1 by a suitable choice of a in A. There is thus a close analogy between this case and making of perfect information in decision-making proper. The only difficulty consists in determining the 'reference point'.

Generally speaking, let p and q be the mixed strategies of players 1 and 2, respectively, that would have been used by them in the ordinary course of events, i.e. if no information about the opponent's strategy was received. Thus, $p(q)$ is a probability distribution over the set $A(B)$. Taken together, p and q determine the outcome in terms of utilities of the players:

$$u_k(p, k) = E_{p(a)} E_{q(b)} u_k(a, b); \; k = 1, 2. \tag{33}$$

The nature of p and q is left undetermined. They may be maximin strategies (in zero-sum games they represent the only reasonable choice), or strategies in equilibrium, or they may depend on some hypotheses concerning the behaviour of the players. At any rate, they determine, through (33), the quantity against which a player's gain or loss, due to the passing of information is appraised.

Once this has been decided upon, it is straightforward to compute player 1's expected gain when he learns about 2's choice, without 2's being aware of it. The value of such information will be denoted by $C_1^2(p, q)$. The subscript refers to the player for whom the change in utility is computed, while the superscript indicates the player whose choice is disclosed to the opponent. The whole expression is, of course, relative to the 'status quo' strategies p and q.

$$C_1^2(p, q) = E_{q(b)} \operatorname*{Max}_a u_1(a, b) - u(p, q).$$

The above formula is easily explained. For any b, player 1 maximizes his utility; the results are averaged by means of 2's strategy q, because it has been assumed that he is unaware of 1's knowledge, therefore his behaviour is not affected.

What if a player does know that his choice will be found out by his opponent? Another way of describing the same situation is: player 1 discloses to his opponent the strategy he has chosen. What does he stand to gain (or lose) by this passing of information?

To simplify matters, we will assume that to any a in A player 2 has only one best reply in B, denoted by $b(a)$.[4] By definition, the strategy $b(a)$ satisfies

$$u_2(a, b(a)) = \operatorname*{Max}_b u_2(a, b(a)). \tag{35}$$

Using this notation, we can easily define the value, for 1, of disclosing his strategy to the player 2. We will use the letter D, other conventions remaining unchanged:

$$D_1^1(p, q) = \operatorname*{Max}_a u_1(a, b(a)) - u_1(p, q). \tag{36}$$

The first term of the above expression is the result of the player 1's maximization under the assumption that 2 will respond to any a in a way that is most advantageous to him.

The value, for 2, of receiving information about 1's choice under the same condition, i.e. with 1's knowledge, is calculated as follows. Let a^* be 1's best strategy[5] in the sense of (36):

$$u_1(a^*, b(a^*)) = \operatorname*{Max}_a u_1(a_2 b(a)). \tag{37}$$

Since player 2 maximizes against a^*, the change in his payoff is

$$D_2^1(p, q) = u_2(a^*, b(a^*)) - u_2(p, q). \tag{38}$$

It is easy to see that while $C_1^2(p, q)$ is non-negative and for p and q, both $D_1^1(p, q)$ and $D_2^1(p, q)$ can assume negative values for some 'natural' status quo strategies such as maximin. As an example, let us consider the following game:

	b_1	b_2
a_1	(3.3)	(0.0)
a_2	(3.2)	(0.4)
a_3	(0.0)	(4.1)

Maximum strategies are: $p_0 = \langle 2/7, 2/7, 3/7 \rangle$. $q_0 = \langle 1/4, 3/4 \rangle$. They determine the points of reference as $u_1(p_0, q_0) = 48/28$ and $u_2(p_0, q_0) = 43/28$. The best replies to 1's declared strategies, together with payoffs they determine, can be tabulated as follows.

		u_1	u_2
a_1	b_1	3	3
a_2	b_2	0	4
a_3	b_3	4	1

We see that 1 ought to declare a_3 which yields 4 as the payoff to him and 1 as the payoff to his opponent. It follows that $D_1^1(p_0, q_0) = 4 - 48/28 = 64/28$, $D_2^1(p_0, q_0) = 1 - 43/28 = -15/28 < 0$. This is an instance of the strategic role information plays in games. For player 2, information of the kind described has negative value: it would be better for him *not* to learn about 1's choice and to apply the maximin strategy instead. That is, if 1 is aware of the passing of information. If he is not, then the value, for 2, of being informed about 1's strategy is, of course, positive: $C_1^1(p_0, q_0) = (3 \cdot 2/7 + 4 \cdot 2/7 + 1 \cdot 3/7) - 43/28 = 25/28$. The corresponding quantity for 1 is $C_1^2(p_0, q_0) = (3 \cdot 1/4 + 4 \cdot 2/7 + 1 \cdot 3/7) - 48/27 = 57/28$. We notice that $C_1^2(p_0, q_0) < D_1^1(p_0, q_0)$. In other words, it is (in this game) more profitable, for 1, to inform 2 of his choice of strategy than to obtain information about the strategy of 2, without 2's knowledge. This 'power of disclosing one's own strategy' has been pointed out long ago in connection with the well-known game, called 'battle of the sexes' (see Luce and Raiffa [1]).

Since different relations may exist between the quantities so far defined, we may ask for the structural properties of the game that entail them. The problems thus created become even more complex when the possibility of misinformation is introduced, i.e. when it is no longer assumed that the strategy declared is identical with the strategy adopted. We are not going to analyze here this problem.

On the other hand, for the special case of zero-sum games the above concepts simplify and we obtain such intuitively appealing relations as, for instance,

$$C_1^1(p_0, q_0) \le D_1^1(p_0, q_0) \le 0 \le D_1^2(p_0, q_0) \le C_1^2(p_0, q_0), \tag{39}$$

where p_0, q_0, are, of course, maximin strategies of the players.

7 SUMMARY

To be informed means: to have a question answered. Our analysis began, therefore, with a typology of questions. In terms of decision-making, a consequence of having information is that one is able to act upon it, in the sense of making one's decision depend on the answer received. The pragmatic value of information is defined in terms of the change in utility or loss, due to the optimal use of information – in the sense of some criterion.

We first analyzed the value of information in decision-making proper, i.e. when the outcome of the decision is co-determined by an impersonal factor. Then the value of information in a conflict with an intelligent opponent was defined for two cases: (i) when the subject learns about the opponent's decision without the latter's knowledge; (ii) when he is able to inform the opponent about his own decision. It was shown that the second type of information has not necessarily positive value; on the other hand, it can be more profitable for the player to give information than to receive it.

Generally speaking, information transforms the decision problem (or game) into another one. In the case of games, the result of this transformation depends on the complex relations of the type 'he knows that I know he knows . . .'. Some consequences of this fact have been indicated above.

NOTES

[1] Sometimes it is also called 'pragmatic information' (cf, K. Szaniawski [10]).

[2] For a more detailed discussion, see K. Szaniawski [6].

[3] More on this topic in K. Szaniawski [11].

[4] Generally speaking, there will be a non-empty set $B(a)$ of best replicas to a, containing possibly more than one element. For a formal description of this case, see K. Szaniawski [11].

[5] Here, again, we assume uniqueness. In general, we would have to speak of a non-empty set strategies satisfying (37).

REFERENCES

[1] Luce R. D., Raiffa H., *Games and Decisions*, J. Wiley, New York 1957.

[2] Marschak J. *Information, Decision and the Scientist*, [in:] C. Cherry (ed.), *Pragmatic Aspects of Human Communication*, D. Reidel Publishing Co., Dordrecht and Boston 1974, pp. 145–178.

[3] McGuire C. B., *Comparisons of Informations Structures*, in: C.B. McGuire, R. Radner (eds.), *Decision and Organization*, Volume in Honor of Jacob Marschak. North-Holland, Amsterdam and London 1972, pp. 101–130.

[4] Ponssard J. P., *On the Concept of the Value of Information in Competitive Situations*, Ecole Polytechnique – CESMAP, 1973, (mimeograph).

[5] Ponssard J. P., Zamir S., 'Zero-Sum Sequential Games with Incomplete Information', *International Journal of Game Theory*, 2 (1973).

[6] Szaniawski K., 'The Value of Perfect Information', *Synthese*, 17 (1967), 408–424. [see this volume, pp. 140–153.]

[7] Szaniawski K., 'On Pragmatic Value of Information', [in Polish:] in J. Kozielecki (ed.), *Problemy psychologii matematycznej*, Polish Scientific Publishers, Warsaw 1971, pp. 325–347.

[8] Szaniawski K., 'Questions and their Pragmatic Value', in: R. J. Bogdan, I. Niiniluoto (eds.), *Logic, Language and Probability*, D. Reidel Publ. Co, Dordrecht 1973, pp. 121–123. [see this volume, pp. 154–156.]

[9] Szaniawski K., 'Information and Decision as Tools of the Philosophy of Science', *Danish Yearbook of Philosophy*, 10 (1973), 47–59. [see this volume, pp. 8–18.]

[10] Szaniawski K., 'Two Concepts of Information', *Theory and Decision*, 5 (1974), 9–21. [see this volume, pp. 157–166.]

[11] Szaniawski K., 'On the Value of Information in Games', in: *Théorie de la décision et applications*, CNIPE Edition, Paris 1975, pp. 37–44.

21

DECISION-MAKING AND FUTURE RESEARCH SOME THEORETICAL PROBLEMS

The remarks below are an attempt to systematize the relations that exist between decision making and the study of future. Also, they list some open problems that arise in this connection.

I shall begin by looking at these relations from the point of view of decision problem, i.e. by asking the question, how the results of future research can (or ought to) influence the decision made at the present time. Then I am going to take as my starting point a classification of predictions, in order to show their formal affinity to the structure of decision. Articulation of certain methodological problems will close this paper, intended primarily as a preliminary statement for discussion.

FROM DECISION PROBLEM TO PREDICTION

There is hardly any doubt that if the word future is taken in a sense broad enough, all decision problems, whether trivial or important, presuppose some ideas of what the future will be, or is likely to be, or at least can be. The reason is rather obvious. Any rational decision ought to be governed by the consequences of contemplated actions that jointly constitute the decision problem. Such consequences belong to the future, relative to the moment of decision making.

Commonsensical as it is, this fact is sometimes obscured by the abstract formulation of the decision problem. Let me discuss it in more detail. In order to do this, I shall have to recall the basic facts about, the conceptual apparatus of decision theory.

According to the general theory of decision-making, a conscious choice of behaviour is determined by at least three factors. The decision-maker must know and be guided by: (1) what he can do; (2) what the relevant features of the environment can be; (3) how he evaluates any combination of his behaviour with the state of the environment. Formally, this approach leads to the definition of a decision problem D as an ordered triple: $D = \langle A, S, u \rangle$, where: A = the set of possible courses of action; S = the set of possible states of the environment (so-called 'states of the world'); u = a real-valued function defined on the Cartesian product $A \times S$ and representing the decision maker's valuations (so-called 'utility').

The decision consists in choosing an element out of the set A. All the uncertainty concerning the consequences of such choice is located in the S. If S has more than one element (i.e. more than one state of the environment is possible) then the consequences of any a in A are not uniquely determined. The valuation, accordingly, is a function $u(a, s)$, where only the first argument is under the decision-maker's control.

The problem D can be enriched by a more detailed characterization of the uncertainty just mentioned. One way to do this is to add a probability distribution p on S. A more sophisticated (and also more general) approach, due to Richard Jeffrey,[1] consists in making the probabilities of states of the world depend on the action adopted. They are, thus, represented as $p(s \mid a)$. Of course, for any $a \in A$: $\sum_s p(s \mid a) = 1$ since whatever the decision maker chooses to do, exactly one state of the world obtains (the elements of S are pairwise exclusive and jointly exhaustive). This approach reduces to the simpler one, mentioned above, if it is assumed that $p(s \mid a) = p(S \mid a')$, for any $s \in S$ and $a, a' \in A$.

A further extension introduces the concept of information on S, making it possible to reduce the uncertainty and thereby improve the decision. I shall return to this point later on. Also, I am postponing the discussion of criteria the theory recommends for the decision-maker to use in the situation given by D.

The above description has been made in a telegraphic style and leaves several serious questions unanswered. Probably the most important one is, how to deal with the possibility of discrepancies between D and the reality it purports to describe. For instance, the set A of actions the decision-maker thinks available may be inadequately defined in at least two ways: it may contain actions which, for one reason or another, are in fact impossible to put into practice; one the other hand, it may not include some effective courses of action (we all know from our own experience this feeling of belated regret: if only I had thought of such and such a possibility . . .). The moral to be drawn is that we cannot be content to accept D as given. Since D can involve serious errors, we shall have to investigate, how D itself (and not its solution) is arrived at, so that some ways of safeguarding against errors can be introduced. This, of course, means going beyond decision theory, which conveniently starts with assuming D to be given and correct.

After this cursory introduction, let us come back to our main question: in what way do the results of future research enter the problem of decision?

One feature of D is striking in this connection: in D, no explicit mention is made of the time parameter. If so, how the consideration of future enters the decision problem at all?

It is, of course, introduced indirectly, in more ways than one. First, the independent factor s (state of the world) may be interpreted as belonging to the future and determining consequences of the actions in A still farther removed in time. More generally, the set S may be thought of as the Cartesian product $S_1 \times S_2 \times \ldots \times S_n$, where S_i is the set of possible states of the world at time i, and the consequences are evaluated at time not earlier than $n + 1$. In other words, any element if S describes the behaviour in time of the uncontrollable factor, up to the moment n. The relevant study of future is concerned with finding out how the environment can develop, i.e. with defining the elements of S. Its next task is to determinate the probabilities of the alternatives constituting S.

This approach seems to be particulary suitable when the interplay between decision-maker's actions and the environment is negligible. They jointly determine the future, but acting in a relatively independent way. Suppose, however, that the environment is influenced by the present decision; *after* this influence has taken effect, the environment co-determines the outcome of that decision. A classical, although somewhat gloomy example (due to Jeffrey) concerns the decision of a state whether to arm or to disarm. The consequences depend, of course, on whether there will be war or not. But the probability of war is not the same in the event of disarmament as in that of armament (the fact has been noticed long ago; cf. the saying: *si vis pacem, para bellum*). It thus turns out that, other elements of D remaining unchanged, the probability distribution on S is a function of a.

This is precisely the case we have mentioned above, when the distribution function p was defined as conditional upon a. A natural generalization would consist in treating the decision problem D as a kind of extensive game with the environment. Both the elements of A and the elements of S would then have to be interpreted as strategies in the game-theoretic sense of the world, i.e. instructions on how to react in any situation that may arise in the course of 'playing the game'. It is well known that such as approach, although theoretically very appealing, leads very quickly to unmanageable complexities, since the number of combinations increases rapidly. Also, it presupposes a rich and accurate knowledge of all possible developments. It thus seems that we must be content with a very approximate description of the interaction between the elements of A and S involving a few alternating moves at the most.

Whichever way is adopted to represent formally the fact that some idea of the future is presupposed in any decision problem, one thing remains constant: the evaluation of present-day decisions in terms of the function u (possibly depending also on some other factors) is determined by the consequences located in some future moment of time. It is characteristic of the state of our civilization, that for most significant decisions this moment of time is rather remote, relative to the time at which the decision is made. Typically, the time distance is of the order of a few decades.

Indeed, it is the urgent need to obtain rational decisions that both explains and justifies the study of future. Mere curiosity, although legitimate in itself, is not enough to provide justification for the expenditure of effort and money on finding out what the future will be like.

The phenomenon (i.e. future research, conducted systematically and on a large scale) is relatively new. The reason for that can be seen in the nature of present-day decisions. Until quite recently, the level of complexity and the dimensions of human actions were limited, in the sense that their undesirable by-products could be either suffered or removed *ex post*. Technical means, now available, have created the possibility that the whole system will be thrown out of equilibrium, beyond the point where that equilibrium could be regained without resorting to drastic measures. We cannot, any more, count on the homeostatic properties of the system: there is no guarantee that it will automatically correct disturbances, pushing its own parameters back into the interval of safety. This is true, whether by 'system' we mean a sufficiently large human community plus its environment, or a state, or the earth and its population taken as a whole.

The danger is the more acute, because of the superposition of a great number of small, independent decisions: their joint effect may easily become catastrophic, as we by now well know. Hence the necessity of acting so as not be taken by surprise. The decision to prevent disaster (possibly by imposing some kind of cooperation) is a particular type of decision. Its necessary condition is reliable knowledge *ex ante*.

To summarize the above argument. There is nothing new in the fact that knowledge of future is needed in order to improve decision, or even adequately to define the decision problem itself. What is new is the concern with non-immediate consequences, particulary the consequences of non-intervention, consisting in accumulated results of uncontrolled development. The role of prediction in decision-making finds no direct expression in the formal structure of decision theory, at least as far as the general concepts are concerned. The theory is synchronic, in the sense that the time variable does not belong to the explicit definition of decision problem. Indirectly, the idea of future influences the evaluation of the present opinions in terms of 'utility' u. This is probably not quite satisfactory and calls for a revision of the traditional apparatus.

FROM PREDICTION TO DECISION PROBLEM

Let us now reverse the situation and attempt to describe the influence of today's decision on the outlook of future. A sizable proportion of predictions must take into account the fact that our present activities co-determine the future; in many cases they are the decisive factor.

Speaking more generally, the object of prognosis may depend (or not) on either the decision, or some uncontrollable factor, or both. Which provides the basis for a natural classification of prognoses.[2] Let Z be the set of possible states of future the prognosis refers to. Let the symbols A and S be interpreted as before: A = the set of available actions, S = the set of possible states of environment, present and uncontrollable. In terms of such a description, there are four possibilities which can be tabulated as follows:

		Does Z depend on S?	
		No	Yes
Does Z depend on A?	No	z	$f\colon S \to Z$
	Yes	$g\colon A \to Z$	$h\colon A \times S \to Z$

The reading of the above table is simple enough. It is presupposed here that the relation between the predicted and conditioning events has a deterministic character. Thus, for instance, prognosis of the type represented in the lower-left corner consists in a conjunction of statements: 'if the action a is taken than z', where a runs over the set A. In other words, the prognosis says explicitly how the future is conditioned by the present decision; the relevant function from A to Z is denoted by g. Formally, prognosis of that type can be identified with g. Analogously, for the remaining cases. Let us say a few words about each of them.

Prognosis of the type z. This is not the only case of unconditional statement about future. Prognoses having this character are, admittedly, rather rare (let me remind the reader that the future spoken about is at least several years removed from the time the prediction is made). Categorical statements of this kind presuppose a good isolation of the system and/or exhaustive and reliable knowledge of its working. A familiar example: predictions concerning astronomical events, such as eclipse of the sun. In the domain of social phenomena, they appear practically impossible.

Prognosis of the type $f\colon S \to Z$. It is a well-known kind of prediction, often referred to as 'many-variant prognosis'. It is applied in the case when the uncertainty about the future can be completely reduced to the indeterminacy of s. If it were known with certainty which of the conditioning events s actually obtains, the prognosis in question would assume the form z, just spoken about. Typically, s intervenes in time between the prognosis and its realization.

Prognosis of type $g\colon A \to Z$. Complete control of future (under a given aspect, in a given time and place) seems to be assumed in this case: the choice of a in A determines uniquely z in Z. Let us notice, however, that if valuations of z were introduced, the control might easily turn out to be illusory, in the sense that none of the events in Z would be judged acceptable. Hence, prognosis of this type very often leads to a revision of the set A: other alternatives are looked for, that would have more satisfactory consequences.

Prognosis of the type $h\colon A \times S \to Z$. This is, in most cases, the only realistic kind of prediction. It assumes that the future is only partially controllable; that the decision-maker's course of action has some influence on what will happen, but is not decisive in the sense of unique determination. Formally, the case under discussion exhibits a marked similarity to the decision problem D. The only step that is needed in order to identify the structure of the prognosis

h with the decision problem D is to introduce valuations, in the form of utility function u, defined on Z. The function u will, in general, depend also on a: the actions are differentiated as to their cost, financially and otherwise.

According to generally accepted opinion, here is the dividing line that ought to be drawn between the responsibilities of the decision-maker and those of the expert. It is not the expert's task to determine the goal, expressed by means of the function u. What is expected of him is a description of the situation, i.e. answer to the question, how the future depends upon the present. The function $z = h(a, s)$ summarizes this description.

Plausible as it appears, the above view is guilty of serious oversimplification. It is, in fact, not so easy to disassociate valuations from the purely descriptive structure h. The definition of the set Z itself (hence also that of A and S) presupposes some accepted values, since only those aspects of the future are taken into account which are not axiologically neutral. The expert, i.e. the author of prognosis, must therefore be aware of the values in question. If they are supplied by the decision-maker, then the last one cannot avoid taking part in the preparation of prognosis.

Usually, this participation is effected through a repeated use of feed-back: the goal, outlined by the decision-maker, leads to a tentative structure h, which causes some adjustment of valuations, which in turn, etc. The procedure is certainly more complex than a simple superposition of values upon a purely descriptive structure of conditional statements about the future, provided by the expert.

To conclude this section, a word of comment on the absence of probability in the above classification of predictions. At first glance, this may seem puzzling, since the existing knowledge is often not sufficient to offer predictions of a categorical nature. The answer is rather obvious: a parallel table, using probabilistic language, can easily be constructed as follows

		Does Z depend on S?	
		No	Yes
Does Z depend on A?	No	$p(z)$	$p(z \mid s)$
	Yes	$p(z \mid a)$	$p(z \mid a \cap s)$

The entries is the above table are, of course, conditional probability distributions. Thus, for instance, $p\,(z \mid a \cap s)$ is the probability of z, given a and s: for any a and s: $\sum_z p(z \mid a \cap s) = 1$.

A natural question is, what kind of probabilities are those used in prediction. To interpret p as (conditional) frequency of z would make it inoperative in most cases, since it is doubtful what the reference set could be and how the frequencies in it are to be computed. An alternative solution is to interpret $p(z)$

as the subjective, or personal, probability of the expert, i.e. his degree of confidence in the prediction that z will happen (possible under given conditions). The problems this interpretation creates are easy to notice. Some of them will be mentioned below.

SOME OPEN QUESTIONS

Let us begin with the problem just mentioned, viz. that of the meaning of probability in prognoses. As we know, it is rather seldom that the probabilities in question are supplied by statistical data, relevant to the determination of comparatively distant future by present events. Statements about future in terms of probabilities express, therefore, the subjective judgment of their authors. They define the odds at which the expert is 'willing to bet' on the future event to which the probability is attached. Two problems immediately arise: (a) are the probabilities genuine, in the sense that they represent an honest and considered judgment of the expert? (b) assuming an affirmative answer to (a), is it safe to act upon them? The expert may be mistaken. How to find this out?

Admittedly, such difficulties are not limited to probabilistic statements about future. Possible dishonesty, or hasty judgment, or just plain error are always something to be reckoned with. The trouble, however, with subjective probability is that it is essentially untestable. Even if the expert tries to justify his judgment by some kind of argument from the testable knowledge he has, he cannot derive numerical probabilities in a unique way. Before such probabilities are stated, there must be a 'jump' for which he alone is responsible. Scientific knowledge is public. Subjective probabilities, by definition, are not. Their conflict (or agreement) with what can be counted as established is not detectable, except in certain extreme cases.

The role of subjective probabilities in the articulation of results of future research explains (at least, partly) the distrust with which such research is looked upon. Long-term probabilistic predictions are virtually immune to falsification. For this reason, they clash with the prevailing view of science. On the other hand, they are simply indispensable as a guide for action. We have already discussed their necessity.

A similar dilemma arises in decision-making, where the most natural criterion for choosing an action in A is that of maximization of expected utility (MEU), the expectation being calculated with respect to a subjective probability distribution on S. An undesirable property of this criterion is that it opens the way to all kinds of self-delusion (wishful thinking, for instance); also, it almost completely rules out the possibility of ascertaining whether the decision has been correct. In cases when the responsibility for decision is to be taken, this is of vital importance. MEU is, therefore, replaced with other criteria, objective in the sense that they make no use of personal probabilities. One instance is the well known principle of maximin. The arbitrariness in the choice of criterion itself is the price to be paid for 'objectivization' of decision.

In the case of future research, such relatively easy solution is not available. there are attempts[3] to solve problem (a), by providing the expert with suitable motivation to do his best. This is effected by means of a kind of game in which the forecaster (assumed not to be the same person as the valuer of outcomes) has an 'accuracy incentive', depending on the true and the predicted outcomes. But this, again, presupposes that the outcome becomes known within a reasonably short period of time. Hence, in the case of long-term prediction the device does not work. The problem of how to objectivize probabilistic predictions remains open. Improvements are probably to be sought in making the basis of prediction as explicit, and therefore as accessible to verification, as possible. With the increasingly hermetic character of expert knowledge, this is not an easy task.

What about raising the level of accuracy of prediction, by means of research, resulting in additional information? In (standard) decision theory, information is assumed to reduce uncertainty concerning the uncontrollable factor S. The idea can roughly be described as follows. If we denote by X the set of possible results of research, the connection between X and S is, generally speaking, probabilistic. Let $p(s/x)$ be, for any given x in X, the probability distribution on S, given x. Then the optimal use of a particular information x in a decision problem D consists in choosing a so as to maximize to average utility, relative to conditional probability distribution $p(s/x)$. The improvement of decision, due to such information, is the difference between the utility obtainable in the presence of information and that obtainable when information is not available (the difference is, of course, averaged over the set X).[4]

Can we apply this argument to the improvement of prognosis? We can, provided that the two factors influencing the future, A and S, are independent. Suppose, in particular, that the prognosis consists in specifying probability distribution of the future events in Z, conditional upon earlier possibilities s: $p(z \mid s)$, for all s in S (upper right corner of the table above). It is implied that those probabilities remain the same, whichever course of action in A is adopted. Then, by rules of probability calculus, any information x influences through $p(s \mid x)$ the prognosis concerning z. The gain in accuracy of prediction can be calculated either in terms of decreased entropy in Z, or in terms of weights attached to different kinds of error.

Obviously, this way is not open when the influence (on the future events z) of the two factors A and S cannot be separated into two parts. In other words, when the decisions and the environment do interact, relative to Z. In such case, it makes no sense to speak of information supplied by empirical research on S, without assuming definite behaviour on the part of the decision maker (nonintervention is, of course, a special type of action). Once, however, a particular behaviour is chosen and implemented, the decision problem ceases to exist and the prognosis becomes pointless with respect to that purpose.

Here is where simulation comes to aid. If the decision cannot be repeated in order to make possible the research on environment co-determining the future, it is sometimes possible to try out different variants on a model showing sufficient similarity with the real system. Computer programmes are devised for this purpose, and simulation games are played. The prognostic value of results thus obtained is, of course, proportional to reliability of the hypothetically assumed analogy and, also, to the stability of conditions.

Can there be some other way of feeding information into the formal description of a system with fully interacting A and S variables? There is, at present, no basis to answer this question categorically.

One final remark to close this brief survey. Once we agree that probabilistic statements about future are essentially subjective, we face the problem of how to deal rationally with conflicting prognoses of this type. Methods of aggregating several subjective judgment (with not necessarily equal weights) into one signal basis for decision are known to exist; one example is the so-called Delphi method, in its many variants. Whatever the refinements introduced, the methods are of a socio-psychological, rather than logical, nature.

Similar problems arise and are dealt with in other fields: the problems are e.g. to reduce many-dimensional valuation to a single hierarchy; to obtain linear ordering of actions under uncertainty as to the state of the environment; to aggregate preferences of members of group into a single 'social' preference. In all such cases, the role of logic consists in the elimination of unreasonable solutions, the meaning of 'reasonable' being defined axiomatically. No unique solution exists in general, sometimes there is no guarantee that the postulates defining the field of acceptable solutions do not come into conflict (cf. the famous 'paradox of democracy', due to Arrow). The problem, how to aggregate subjective estimates by forecasters, belongs to this formal group.

NOTES

[1] Cf. Richard Jeffrey, *The Logic of Decision*, McGraw-Hill, 1965.

[2] I am here following closely the argument in K. Szaniawski, "Prognoza a podejmowanie decyzji" [Prognosis and Decision-Making], in: *Studia z teorii i metodologii prognozowania społecznego* [Studies in Theory and Methodology of Social Prognosis], ed. by A. Siciński and A. Rażniewski, PWN, 1976.

[3] Cf. I.J. Good, "Information, Rewards and Quasi-Utilities", in: *Science, Decision and Value*, ed. by J. Leach, R. Butts and G. Pearce, D. Reidel Publ. Co, 1973.

[4] A more detailed description may be found in K. Szaniawski: "Two Concepts of Information", *Theory and Decision* 5, 1974, 9–21 [see this volume, pp. 157–166.].

22
ON FORMAL ASPECTS OF DISTRIBUTIVE JUSTICE

1 INTRODUCTION

The concept of distributive justice is here considered in an extremely simplified form, viz. relative to the situation when n objects are to be divided among m persons ($1 \leq n \leq m \geq 2$) and no account is taken of factors that could differentiate the persons as far as this division is concerned, such as their merits or needs. By so limiting the concept of justice, we gain the possibility of a formal analysis. On the other hand, the applicability of such a concept in actual social contexts is rather probabilistic.

In the model under consideration, distributive justice essentially amounts to some form of symmetry. Probably the most natural way to express this symmetry is to postulate that everybody has the same chance of getting what he desires. The problem then arises, whether such postulate is consistent with another natural desideratum, viz. that the division of objects between individuals be Pareto optimal. This question must be answered in the negative. It has been shown[1] that the two postulates are, in general, incompatible.

The present essay is an attempt to push the analysis a little further. In particular, I will try to show that 'postulate of equal chances' can be replaced by another postulate which does not contradict the Pareto optimality. It thus appears that even in so simple a model, we can have more than one concept of justice. Whether the two concepts are equally satisfactory, is a matter of ethical judgement.

I shall begin by restating the problem and the earlier results. The alternative postulate will then be formulated and compared with the previous one.

2 BASIC CONCEPTS

Let $G = \{A_1, A_2, \ldots A_n\}$ be a set of objects which are to be distributed among individuals belonging to the set $P = \{1, 2, \ldots, m\}$. The word 'object' is given here a wide interpretation. For instance, the right to sit in the King's presence may also be called object. However, unlike the last-mentioned privilege, elements of G are treated as unique: no object may go to more than one person.

For simplicity, elements of G will be denoted by A, B, C, ... It is assumed that there are at most as many objects as persons: $n \leq m$. The reasons for this restriction will be given below. Let $\leq_k$ be the (weak) preference ordering of the set G by the person k. Strong preference $<_k$ and indifference $=_k$ are defined in the usual way. The set $\{\leq_i, \leq_2, \ldots, \leq_m\}$ is called 'profile of preference orderings'.

The empty set $\varnothing$ is also subject to evaluation in this sense, since one must allow for the possibility of someone's getting no objects from the set G: we shall then say that this person gets $\varnothing$. It is postulated that

$$\text{for all } k \text{ and all } j\text{: } \varnothing <_k A_j. \tag{1}$$

In other words, everybody strongly prefers getting any object to getting nothing. In this sense, elements of G are 'goods'.

By *division* I shall mean any way of attributing objects to individuals, i.e., any function from G to P.

Very little can be said about divisions before further concepts are introduced. It is, however, clean that in view of $n \leq m$ and (1), a division attributing to some person more than one objects would violate the principle of justice. A postulate[2] eliminating such divisions will be adopted:

No person gets more than one object. $\quad (R)$

From now on, I shall consider only divisions satisfying R.

Further analysis is made possible by introducing randomization. Its role is, of course, to 'smooth out' inequalities inherent in necessarily discrete divisions. In the simplest case, when two people want to get one indivisible object, the ultimate inequality is unavoidable. Symmetry is then saved by giving them equal chances of getting what they want. It is, indeed, a common practice to 'toss a coin for it'. The definitions below are intended as a generalization of this simple solution.

Any probability distribution on the set D of all divisions will be called *distribution*. Division is, of course, a special case of distribution in this sense.

A *distribution rule* is a function from the set of all profiles of preference orderings to the set of all distributions. In other words, a distribution rule determines, for each profile of preference orderings, the random procedure to be followed in the choice of actual division. The central problem now is, how to define distribution rules that are ethically satisfactory?

3 EQUAL CHANCES OF SATISFACTION

In the absence of any other criteria differentiating the participants in the division, distributive justice must consist in treating them on an equal basis. Perfect equality in the actual division of goods is, in general, unattainable. We can, however, postulate equality in terms of chances. It is not accidental that, in the trivial example above, *fair* coin was used to decide who will get the desired object. Any other distribution would violate the principle of equal treatment of the two participants.

In order to generalize this idea, let the objects in G be numbered, for each individual, according to their preference-ordering by that individual. (We assume, from now on, that indifference does not occur. The assumption entails no loss of generality, since indifference makes it easier to solve the problem: if someone is indifferent between A and B, his tastes can be accommodated in more ways than when he strongly prefers A to B.) Let $p_i k$ denote the probability that individual i gets the k-th objects in his hierarchy of preferences. It is natural to adopt the following postulate.

$$\text{For all } i, j = 1, \ldots, m \text{ and all } k = 1, \ldots, n : p_i(k) = p_j(k). \tag{E}$$

In the other words, the distribution rule satisfying E must be such that everybody has the same probability of getting the k-th object in his preference ordering ($k = 1, \ldots, n$).

Let us see how E works in some specific case. In order to avoid clumsy formulations, I will limit the analysis to the case $m = n$. The limitation leaves unchanged the conceptual structure of the problem.

In view of R, each person gets exactly one object. Any division can, therefore, be represented as a permutation of elements in G, under the interpretation that i-th element in the permutation is assigned to the i-th person. For $n = 3$, we have 6 possible divisions, an easily manageable number. In the discussion below, I will use the following examples.

(a)	(b)	(c)	(d)	(e)
1 2 3	1 2 3	1 2 3	1 2 3	1 2 3
A A A	A B C	A A A	A C C	A A C
B B B	...	B B C	B B A	B B B
C C C	...	C C B	C A B	C C A

Each case represents the structure of preferences in the 3-elementary group. Objects are listed according to their decreasing order of preference in a given person's hierarchy. Thus, in case (d), the person 3 wants C in the first place, then A, then B.

All possible divisions, together with their probabilities x_i, are:

$$\begin{array}{ll} ABCx_1 & BCAx_4 \\ ACBx_2 & CABx_5 \\ BACx_3 & CBAx_6 \end{array}$$

Given the profile (c) of preference orderings, the postulate E imposes on the distribution $\{x_i\}$ a condition expressed by the following set of equations.

$$\begin{aligned} x_1 + x_2 &= x_3 + x_4 = x_5 + x_6 \\ x_3 + x_4 &= x_1 + x_6 = x_1 + x_3 \\ x_5 + x_6 &= x_2 + x_4 = x_2 + x_5 \end{aligned} \tag{2}$$

For instance, the equations in the second line ensure that 1, 2 and 3 have the same probability of getting the object that stands second in their hierarchy of preferences, defined by (c). For 1, this object is B. B is assigned to 1 in the third and fourth division. Therefore, $x_3 + x_4$ is the probability that 1 gets the second object of his preference. By the same reasoning, this probability is $x_1 + x_6$ for 2 and $x_1 + x_3$ for 3.

The set (2) is, of course, redundant and can be reduced to:

$$\begin{aligned} x_1 &= x_4 = x_5 \\ x_2 &= x_3 = x_6 \end{aligned} \tag{3}$$

An insight into the meaning of these conditions is obtained when divisions are represented by the ordinal numbers of objects assigned to the respective person:

$$\begin{array}{ll} 122\ x_1 & 133\ x_2 \\ 231\ x_4 & 212\ x_3 \\ 313\ x_5 & 321\ x_6 \end{array} \tag{4}$$

It is then easy to see that the probabilities of obtaining objects ranking as No 1, No 2 and No 3 are, for 1, 2 and 3:

	1	2	3
No 1	1/3	1/3	1/3
No 2	1/3	1/3	1/3
No 3	1/3	1/3	1/3

(5)

This follows from (3) and (4). In each column of such a table the probabilities must of course sum up to 1.

A distribution satisfying E exists, for all profiles of preference orderings. It simply is the equidistribution $x_i = 1/n!$. When all divisions are equiprobable, the probability that the j-th individual gets the objets Ńo 1 in his preference ranking is $(n-1)!/n! = 1/n$. And the same argument goes for objects No 2, etc. Therefore, in tables of the type (5) all the entries are equal to $1/n$, which shows that the postulate E is met by such a distribution.

4 OPTIMALITY

The notation adopted in (4) suggest another desideratum, besides that of distributive justice. When we compare, for instance, the first two divisions (to denote them, I will use the symbols d_i), it easy to notice that d_1 dominates d_2, in game-theoretic sense. Under d_1, each person gets at least as much as under d_2, while persons 2 and 3 actually get more, since the objects assigned to them rank as second instead of third. Obviously, d_2 is an unsatisfactory way to assign objects to persons, as long as d_1 is possible. According to the principle of Pareto optimality, we ought to exclude such divisions as inadmissible.

Formally, let a division be represented as $k_1, k_2 \ldots k_n$, where k_i is the rank number of the objects assigned to the person i. We shall say that $k_1, k_2 \ldots k_n$ dominates $l_1 l_2 \ldots l_n$ when $k_i \leq l_i$ for all i and $k_i < l_i$ for some i. The postulate of Pareto optimality can be expressed as:

> If a division is dominated by some other division its probability is zero. (*O*)

Out of all possible variants of optimality, that of Pareto seems to be the weakest, therefore the least controversial. Also, the requirement above is consistent, in the sense that it can be met for all possible profiles of preference orderings. This follows from the fact that the relation of dominance is transitive. In a finite set of divisions, partially ordered by that relation, there must be at least one non-dominated element. The postulate *E* and *O*, however, turn out to be inconsistent.

5 EQUALITY VERSUS OPTIMALITY

Their inconsistency is exhibited in the example (*c*), discussed above. By the postulate *O*, $x_2 = 0$ since d_2 is dominated by d_1, and $x_5 = 0$ since d_5 is dominated by d_3. By the postulate *E*, $x_2 = x_3 = x_6$ and $x_5 = x_4 = x_1$. It follows that $x_i = 0$ for $i = 1, \ldots, 6$. Which means that a distribution satisfying *E* and *O* for preference orderings described in (*c*) does not exist. But a distribution rule associates a distribution with *each* profile of preference orderings. Therefore, a distribution rule satisfying both *E* and *O* does not exist.

When faced by inconsistency of simple and intuitively convincing requirements, we can look for two types of solution. The first one consists in weakening of one or more of the desiderata. In the present case this seems hardly feasible. The Pareto optimality is already an extremely weak version of optimality. As to the equality of chances, one could of course restrict it to objects ranking as first in the respective hierarchies, but the intuitively convincing character of the postulate *E* would then be lost.

Another way out is analogous to that followed in the case of the Arrow postulates for democratic welfare function. It was shown that their inconsistency vanishes when the range of their application is suitably limited. The question is, for which profiles of preference orderings the postulates *E* and *O* can be jointly satisfied?

It is rather obvious that *E* and *O* are consistent when the conflict of interests between the participant in the division is either maximal or minimal. In the first case, represented by example (*a*), all preference orderings are identical. Domination cannot arise, since in any division, an improvement of someone's position implies a worsening in the position of another person (the situation is somehow analogous to that of zero-sum game, where any arbitration scheme fails because of complete opposition of interests). It follows that a dominated division does not exist and postulate *O* is vacuously satisfied. In the example (*b*), there is no conflict of interests: the highest ranking objects are different for each person (the rest of the preference scale is immaterial), therefore the division *ABC* dominates all other divisions. At the same time it satisfies, by itself, the postulate *E*, since it is represented by identical numbers 111. Hence, the distribution assigning probability 1 to *ABC* satisfies both *E* and *O*.

The question, whether the disjunction of (*a*) and (*b*) is a necessary condition for the consistency of *E* and *O*, must be answered in the negative. Cases (*a*) and (*b*) are not the only ones for which the two postulates can be reconciled. Case (*d*) serves as a counter-example. For this structure of preferences, the postulate *E* requires the following grouping:

d_1	ACB	113	$x_1 = 0$
d_2	BAC	231	$x_2 = 0$
d_3	CBA	322	$x_3 = 0$
d_4	ABC	121	$x_4 = 0.5$
d_5	BCA	212	$x_5 = 0.5$
d_6	CAB	333	$x_6 = 0$

(6)

By *E*, we have $x_1 = x_2 = x_3$ and $x_4 = x_5$. By *O*, $x_3 = 0$ (d_3 dominated by d_4 and d_5) and $x_6 = 0$ (d_6 dominated by any other division). Therefore, the distribution on the right-hand side is implied side is implied *E* and *O*. It yields the table of probabilities:

	1	2	3
No 1	0.5	0.5	0.5
No 2	0.5	0.5	0.5
No 3	0	0	0

(7)

We thus see that joint satisfiability of *E* and *O* is not limited to the two extreme cases of total conflict. A sufficient and necessary condition is, however, not easy to find.

6 EQUAL CHANCES OF CHOICE

The postulate to treat symmetrically the participants in the division can be interpreted in more ways than one. Instead of speaking in terms of a person getting the *k*-th objects of preference, we could consider the order in which the participants are allowed to make their choice. Equality would then consist in such a randomization that everybody has the same chance of being the *k*-th to make his choice ($k - 1, \ldots, n - 1$; the *n*-th person has, of course, no choice but to take the object that is left; if $n < m$, the same goes for the remaining persons: $n + 1, \ldots, m$ who then get $\varnothing$). Formally we have the postulate

$$\Pr\{i \text{ is the } k\text{-th to choose}\} = \Pr\{j \text{ is the } k\text{-th to choose}\} \qquad (C)$$

for all $i \neq j = 1, \ldots, m$ and all $k = 1, \ldots, n - 1$.

This amounts to postulating equiprobability of all possible orders in which people present themselves to make their choice.

The use of the term 'choice' in the wording of postulate *C* helps to see the rationale of *C*. The 'decisional' language is, however, inessential, since *C* can be expressed directly in terms of individual preferences. The procedure is as follows.

For any permutation of participants in the division, we assign to the first element of that permutation the highest ranking object in his hierarchy of preference; out of the remaining objects, we assign to second person the highest ranking object in his hierarchy of preference; and so on, until the set *G* is exhausted. In this way, we obtain a specific division, determined uniquely by the permutation we started with (the uniqueness depends on the assumption that only strong preferences are allowed). Postulate *C* demands that all permutations of participants be equiprobable. Therefore, all the divisions generated by them must also be equiprobable. Of course, some divisions appear more than once, because they are generated by more than one permutatation. Their probabilities increase accordingly. Let us see how the postulate *C* works in the case of example (*c*), used above to show inconsistency of *E* and *O*. Next to the permutations of the elements of *P*, I list the divisions they generate.

123	*ABC*	231	*BAC*	
132	*ABC*	312	*BCA*	(8)
213	*BAC*	321	*CBA*	

(8) entails the following distribution:

ABC	2/6	*BCA*	1/6	
ACB	0	*CAB*	0	(9)
BAC	2/6	*CBA*	1/6	

The table of probabilities, analogous to (5), summarizes this results:

	1	2	3	
No 1	2/6	2/6	4/6	
No 2	3/6	3/6	2/6	
No 3	1/6	1/6	0	(10)

It is easy to see that the two divisions, *ACB* and *CAB*, responsible for the conflict of postulates *E* and *O* (because they are the dominated ones), now have probability zero, which means that the distribution (9) satisfies Pareto optimality. This is no accident. The procedure employed in the transition from persons to corresponding divisions automatically ensures Pareto optimality, since in each step an individual is assigned his *best* object that are still available. Dominated divisions are thus eliminated and there can be no conflict between postulates *C* and *O*.

A comparison of (10) with (5) shows that equality of chances exhibited in (5) is now violated in favour of the person 3. This phenomenon is easily explained by the fact that 3 profits from his somewhat excentric tastes: preferences of 1 and 2 are identical, hence the competition between them is complete. It is also worth noticing that the position of 1 and 2 is also improved when we pass from (5) to (10). Their gain was made possible by the fact that 'absolute equality', postulated in E and implemented in (5), has been given up.

7 TWO VERSIONS OF EQUALITY

Let me first point out some formal differences between E and C. The postulate C determines uniquely the distribution, for any profile of preference orderings (provided that only strong preferences are allowed). On the other hand, E specifies a class, usually infinite, of distributions that are satisfactory form this point of view. For instance, there are infinitely many distributions satisfying (3), although they all yield the table (5). It follows that E and C diverge when the structure of preferences is such that the C-distribution does not belong to the E-class of distributions.

It is rather obvious that the extreme cases (*a*) and (*b*), i.e., total conflict of interests and complete lack of conflict, do ensure the agreement of E and C. They are, however, not necessary for the agreement to take place. If individual preferences exhibit certain symmetry, this is reflected in the C-distribution which, for this reason, satisfies the requirement E. For instance, in the case of $m = 4$, the profile

$$\begin{array}{cccc} 1 & 2 & 3 & 4 \\ A & A & C & C \\ B & B & D & D \\ D & D & B & B \end{array} \tag{11}$$

yields, by C, equiprobability of the four non-dominated divisions

$$\begin{array}{lll} ABCD & (1212) & 0.25 \\ ABCD & (1221) & 0.25 \\ BACD & (2112) & 0.25 \\ BADC & (2121) & 0.25 \end{array} \tag{12}$$

while any other gets the probability zero. The distribution (12) evidently satisfies E, by granting every person probability 0.5 for the top object and 0.5 for the second best object. The essential difference between the two requirements consists, of course, in the meaning of equality. According to E, the participants are (probabilistically) equal with respect to the degree to which their wishes are satisfied. According to C, they are given (probabilistically) equal access to the set of objects they value. This is far from being the same thing.

It can be said that E demands more, in the sense that it postulates equal satisfaction of tastes, irrespective of what these tastes are. Postulate C is more formal: in demanding equal access to G, it makes the degree of satisfaction depend, for each individual, upon the configuration of tastes in the group.

For some profiles of preferences, C can yield results which seem rather unintuitive, unless one knows the procedure by which they have been obtained. Thus, for the profile (*d*), we obtain by C the following probabilities.

	1	2	3
No 1	4/6	3/6	3/6
No 2	2/6	3/6	2/6
No 3	0	0	1/6

(13)

When we compare this with the symmetric table (7), we see that in the change from *E* to *C*, the first individual has gained and the third has lost, while the position of the second person has remained unchanged. It is not immediately clear why this should be so.

On the other hand, the desirable property of *C* is that the distribution it determines is always Pareto-optimal, i.e. dominated divisions are excluded. The *C*-distribution can be strikingly efficient as shown by the example (*e*). Given such profile of preference orderings, the condition imposed by *E* defines a family of distributions in terms of the numbers a and b $(a + b = 1/3)$, whereas *C* eliminates all divisions except those represented by the ordinal numbers 121 and 211.

ABC	*a*	0.5	*ACB*	*b*	0
BCA	*a*	0	*BAC*	*b*	0.5
CAB	*a*	0	*CBA*	*b*	0

(14)

The resulting tables of probabilities for *E* and *C* are, respectively

	1	2	3
No 1	4/6	3/6	3/6
No 2	2/6	3/6	2/6
No 3	0	0	1/6

	1	2	3
No 1	1/2	1/2	1
No 2	1/2	1/2	0
No 3	0	0	0

(13)

The contrast between fundamental equality and optimal allocation of objects is here apparent.

SUMMARY

There is, of course, much more to the concept of distributive justice than what was discussed above. It is usually assumed that members of the society have a common, or at least prevailing, hierarchy of values. Then the main ingredient of justice is a criterion, according to which the distribution takes place. Perhaps the most difficult problems are due to the conflict of several, equally acceptable, criteria.

In the present analysis, it was assumed that the only factor by which the individuals are differentiated, was the preference ranking of goods. Justice was reduced to egalitarianism. In order to make equality possible, randomization was introduced, in spite of its somewhat artificial character.

It turned out that even in such extremely simple situation there are at least two acceptable concepts of justice. The first one is more radical, in the sense that it postulates equal degree of satisfaction for everybody. It has the drawback of being non-optimal. A more modest requirement is, to give everybody equal opportunity to choose. It implicitly accords privilege to people whose preferences deviate from the prevailing pattern.

Finally, a words of comment on the relation between the problem of distribution rule and that of social welfare function. Similarity of data is obvious: in both cases, the domain of the function is the set of all profiles of preference orderings. Their ranges are different. But the analogy goes further than that. It is easy to notice that, for each individual, the preference ordering of the set G of goods determines uniquely the preference orderings of the set D of all possible divisions. We, therefore, can translate the original problem into the terminology of Arrow. The social welfare function would lead from the set all of profiles of preference orderings of D to D itself. By the impossibility theorem, no such function satisfying the postulates of democracy exist. Unless, of course, the translation from preferences on G to preferences on D restricts the domain of the function in such a way as to make the theorem of Arrow inoperative. This, however, is another problem.

NOTES

[1] K. Szaniawski, 'O pojęciu podziału dóbr' ('On the Concept of Distribution of Goods'), *Studia Filozoficzne* 2 (1966). The main result is described in K. Szaniawski, 'Formal Analysis of Evaluative Concepts', *International Social Science Journal* 27 (1975). [see this volume, pp. 128–139.]

[2] The postulate has formerly been expressed in a more general way, viz. that no one gets $k + 2$ (or more) objects while someone else gets k objects. Coupled with the assumption that there can be more objects than persons, this had counterintuitive consequences. Preference ordering of the set G does not uniquely determine preference ordering of the set 2^G. Thus we can have it that both person 1 and 2 strongly prefer A to $\{B, C, D\}$. Then, under the natural assumption of monotonicity: $X \subset Y \Rightarrow X \prec Y$ (X and Y ranging over 2^G), the division d_1 attributing A to 1 and $\{B, C, D\}$ to 2 is more egalitarian than the division d_2 attributing $\{A, B\}$ to 1 and $\{C, D\}$ to 2. But d_1 is prohibited by the above postulate, while d_2 is not.

One way to block this undesirable consequence is to introduce the restriction: $n \leq m$. The danger in unrestricted use of R was pointed out to me by Mogens Blegvad. I want to thank Professor Blegvad for his valuable criticism.

23

PHILOSOPHY AND DECISION-MAKING*

It is natural to assume that the *philosophy* of decision-making is the *theory* of decision-making, or at least is about the general assumptions of such a theory. Thus one might say that one such philosophy consists in treating decision-making as being the result of maximization; more specifically, maximization of a compound of values and beliefs. Another philosophy tries to make decision-making objective by eliminating, or at least playing down, the role of beliefs (which are interpreted as subjective estimates of chance). There are, of course, several variants of these two philosophies.

Therefore philosophizing about decision-making appears to mean comparing and discussing general assumptions that underlay various theories of decision-making. Since this has been done often enough, I do not see that considering these issues once again would be a very promising enterprise. I would rather focus my attention on two problems:

1. What can thinking about decision-making in general, from a philosophical point of view, teach us about the decision-making process?
2. What kind of influence has the analysis of decision making exercised upon philosophy? That is, how can philosophy profit from the conceptualizations and experimentation of different decision theories?

The first question does not requires any justification, if we are interested in philosophical aspects of decision-making. Besides, I would like to stress that when I put this question my task is purely semantical. I do not take here any attempt to offer a structuralization of decision problems of arbitrary complexity. This last problem can be solved in various ways (recently this was investigated by von Winterfeldt 1980). This second question is probably more interesting for philosophers than to someone who is involved in analysis of decision-making. However, it is obvious that such influences are usually mutual. By a well-known feedback principle, attempts to apply devices of decision theory for formulating and resolving philosophical problems resulted in many improvements of the theory itself. This last fact perhaps sufficiently justifies the second part of my introductory remarks.

THE CONCEPT OF DECISION

In this essay *decision* is treated basically as *the resolution to do something* or, better, *the resolution to behave in a certain way*. 'Behaviour' is defined here very liberally, covering the act of abstaining from action, complex chains of simple actions, 'strategies' in the game-theoretic sense of the word, or even adoption of a purely mental attitude (e.g. acceptance of a certain view). Perhaps the use of the word 'behaviour' in this context is not a very good choice since psychologists are accustomed to interpreting the word as denoting only externally observable states of the individual. I, therefore follow Sjöberg's (1979) suggestion to use term 'action' instead of 'behaviour'.

Secondly, decision means a *choice*. This is only natural, since a resolution to behave in a given way presupposes at least two options (to behave that way or not); it is my opinion that to speak of decision when there is merely one option would violate the received meaning of the word. Thus the minimal connotation of 'decision', in so far as 'choice' is concerned, can be represented formally by the ordered pair $\langle A, a_0 \rangle$, where A is a set having at least two elements and a_0 belongs to A as the selected element. The set A consists of alternatives of which the subject (called S) is aware. In most cases, this is only a subset of the class, say A', of the options actually open to S.

Is A well-defined? Theories of decision-making assume that it is, in the sense that S is able to distinguish clearly between what belongs to A and what does not. Compared with real-life situations of decision-making, however, this seems to be an oversimplified view. Actually, the set A may be fuzzy. This means that S may be hesitant about the feasibility of some elements of A and/or their properties or that he is not quite clear about the goal that is guiding his choice.

One consequence of the fuzziness of A is that S cannot be unconditionally committed to a_0 for it may turn out that in point of fact a_0 is not feasible. If it were feasible, it would be quickly adopted, but at the time of deliberation, its feasibility is not clear. It is quite often that the only way to achieve certainty about the feasibility of a given alternative is to try to implement. If this is too risky, it complicates the problem of choice by imposing upon S the necessity of selecting some other course of action, one less desirable than a_0 but more feasible.

More generally, and perhaps more philosophically, how is A obtained at all? Certainly it does not represent the totality of *literally all* actions available to S, even subjectively, at the time of consideration. This totality is too large to be considered within a reasonable period of time. Also an overwhelming majority of the available actions are not 'eligible' relatively to the specific goal that motivates the decision in the first place. Thus, for example, if S's goal is to get to his university lecture on time, one element of A might be to wait for the bus (hoping that it will arrive on schedule) while another might be to take a taxi (more reliable but more expensive); buying a newspaper or walking in the park are ineligible elements since they do not advance S toward his goal.

It would seem from the above analysis that the standard way of thinking about decision-making is misleading to some extent; it is based on the assumption that valuations are defined for *A* (or the Cartesian product of *A* and the set of states, say *X*, of some independent factor). This suggests that *A* exists prior to the values that guide the choice. In fact, however, the values are crucial in the determination of *A* itself. The goal to be achieved by *S* precedes the definition of *A*, whether *A* is fuzzy or not. Thus the standard way of thinking about decision-making applies only after the preliminary selection of 'eligible' actions (in view of the goal) has been performed.

If the above analysis is correct, then the set *A* is generated by the goal *G* together with the corpus of knowledge *K* which makes it possible to distinguish between those potential actions that are suitably related to *G* and those that are not.[1] What are the features that make this distinction plausible? In the opinion of *S*, an action *a* must for its inclusion in *A*, satisfy two conditions:

1. *A* must be acceptably feasible, that is, the power of *S* is sufficient for performing this action.
2. *A* must be goal-relevant. This may be interpreted in several ways, but formulation in terms of causality is probably the best interpretation.

That is, an action *a* is relevant to the goal *G* iff there is a causal nexus between *a* and *G*. Following a popular conception of causality (Kotarbiński 1955), one could assume that *a* is relevant to the goal *G* iff *a* is necessary element of a sufficient condition for *G*. Both conditions, feasibility and goal relevance, are evaluated on the basis of *K*.

So much for the awkward problem of complexity of *A*. If its elements were not differentiated, i.e., if they were equally good in terms of goal-achievement and were equally valued from other points of view, then the problem of decision as a choice would cease to exist. If the set *A* has been defined, any of its elements would be suitable. However, this case is very rare.

Axiologically the elements of *A* have a number of features that differentiate them in significant ways. Some are intrinsically better or worse than others (e.g., from a moral point of view or in terms of pleasure). Some have side effects which are better or worse than the side effects of others. Some elements appear more likely than others to bring about the goal, or, perhaps, the outcomes that some might produce, aside from *G* or not-*G*, are unclear. Decision theory has traditionally concentrated on this last aspect of the elements of *A* that is the lack of clarity whether they will produce the desired outcomes. This uncertainty problem has been dealt with in several well-known ways, e.g., by assuming that uncertainty can be measured and such measures can be aggregated, etc. However, here I am concerned with what is common to the different approaches, not with what differentiates them.

This reasoning can be summarized in the following diagram:

$$G \xrightarrow[K]{} A \xrightarrow[VK]{} \langle A, \{u_i\} \rangle \xrightarrow[?]{} \langle A, a_0 \rangle$$

The goal G represents the starting point. Together with knowledge K about the feasibility of actions and their causal connections with G, the set A is generated. The body of knowledge K and the accepted standards of value V lead to evaluations u_i of A's elements. This symbolism covers the case in which the plurality of value-evaluation is due to the intervention of some uncontrollable factor, called nature, the environment, etc. The formal nature of u_i's is left unspecified; minimally, they are weak orderings of A.

The crucial step is, of course, the final one which leads to the actual selection of a_0 from A. The resolution consisting in doing a_0 is what goes by name 'decision'. Decision theory tries to capture this last step, symbolized here by a question mark (?). As we are perfectly aware, there is more than one way of doing this: if the purpose of the theory is normative, i.e., to define a rational transition from $\langle A, \{u_i\} \rangle$ to $\langle A, a_0 \rangle$ then the main difficulty, which is philosophically quite interesting, is where we should to draw the line between what is arbitrary and what is necessary in the decision from the point of a view of rationality. Any attempt to define the transition (i.e., ?) must answer the delicate question, 'What restrictions can be 'reasonably' imposed upon the free choice by S of a_0 from A?'

In fact, the process of implementing the decision is much more complex than it would follow from the above diagram. Certainly, it is not a linear process or at least not fully linear, because a feedback occurs in it. The diagram was not constructed in order to reflect this process. One could rather say that the diagram represents the features which seem important for the concept of decision, if there is such non-empty concept at all.

A decision is not only a *choice*; it is also a *commitment*. Since this is the case, one might question whether there can be a discrepancy between what can S choose (in the sense that he believes it to be the best way to act, given A) and what he actually does. This issue is a classical one in the history of philosophy and a 'yes' answer is at least as old as the dictum by Ovid: *Video meliora proboque, deteriora sequor*. According to this philosophy, people are notorious for following what they know to be the worse course, worse not necessarily in the sense of some impersonal (e.g., moral) standard of value, but worse in terms of the person's own interest, as, in fact, it is perceived by him. (This is one possible meaning of the so-called 'irrationality' of human actions.)

The 'no' answer probably originated with the assertion by Socrates that it is enough to *know* what is good in order to *act* accordingly. (This opinion was shared by many philosophers, particularly in of the French Enlightment.) Interpreted as a specification of the sufficient condition for *morally* good behaviour, it presents a somewhat optimistic conception of human motivation; perhaps it would be difficult to take this claim too seriously in the second half of our own century.

The modern analogue of this latter view asserts that it is proper to say that when S implements a_0, it indicates that S considered a_0 as the best alternative available to him. The validity of this statement depends, of course, on whether we adopt a definition that suitably limits the use of the expression 'S feels a_0 to be the best alternative available to him.'. Behaviour would thus become a symptom (according to a behaviourist, *the* symptom) of a decision. Like all linguistic conventions, this one can be arbitrarily adopted or rejected. Whichever it is, there are far-reaching consequences for the status of the rationality assumption in explaining human behaviour. Since I am speaking primarily about the concept of decision in general, I shall not develop the issue further.

In sum, then, taking the narrowest point of view, the concept of a decision includes two elements: the choice of an action and commitment to (i.e., subsequent performance of) the action that is chosen. However, a wider view reveals that we may also include in the decision the action of defining the set A out of which the actual choice is to be made. If we do so, the concept of a decision becomes complexly related to the goal, since it is the goal that specifically determines the set A. The corpus of the individual's beliefs and values is also essential, both in defining A and in influencing the choice, but primarily in a general way since it does not necessarily change from one decision to another. The goal, however, is specifically related to the decision and therefore a broader conception of decision makes its connection with the goal quite explicit.

THE INFLUENCE OF DECISION THEORY ON PHILOSOPHY

The second part of this essay concerns the changes that analyses of decision-making have brought to the way in which classical problems of philosophy are discussed. Three philosophical disciplines will be considered: ethics (including general axiology), epistemology, and praxiology. Since what follows will cover areas that are rather well-known, each discipline will be covered rather briefly.

Ethics

1. The problem of numerical representation of values. As far as I know, this problem has not been taken up by traditional philosophy. Some work on cardinal utility has been done by economists, of course, but in a somewhat *ad hoc* way. The axiomatic treatment of value, including attitudes toward risk, is due to game theory and decision-theoretic interpretation of mathematical statistics. One logically elegant aspect of this (which has become so commonplace that it is difficult even to notice it) is that it adopts preference ordering as both the starting point and the criterion of adequacy of numerical representation. Development and refinement of classical results continue to the present (cf., for example, Jeffrey 1978; Domotor 1978). There is no need here to discuss the differences between, say, the unified treatment of values *and* probabilities

(Jeffrey 1965) and the theories in which they are treated separately (von Neumann and Morgenstern 1944; Savage 1954). However, it should be stressed that the growing dissatisfaction with the 'global utility' approach does not detract from the philosophical significance of attempts to represent values in such a way that they can be mathematically combined. The 'arithmetic of happiness' in the Bentham style was not an encouraging precedent.

2. The choice of the ethically best alternative in the face of uncertainty. *Prima facie*, this is just a special case of the general problem of decision under uncertainty; specialization is effected by interpreting the value ingredient of the general problem as ethical value. A subtler approach has been taken by Suppes (1973). By explicitly introducing the concept of moral obligation, he has been able to show the connections with certain formulas of deontic logic. Apparently this line has not been continued, so there is not much else to be said about it.

3. Formal representation of conflict of values. It was noticed early on that in the familiar utility matrix, where rows correspond to actions, the columns could be interpreted as valuations from different points of view by the same individual (instead of valuation under different possible states of the world). A solution of the conflict of values would then consist in aggregating the columns into a single column with the maximal element as the 'globally best' action.

Unfortunately, solutions advocated for the standard (uncertainty) interpretation do not automatically carry over to the aggregation of values. For example, uncertainty can, in principle at least, be diminished or even eliminated altogether by means of additional information. However, no information can make a genuine conflict of values less acute, unless, of course, we attribute such conflict to imperfect self-knowledge which sometimes is indeed the case.

What decision theory has to offer towards the solution of conflict of values does no go far beyond simply expressing the problem itself in a more exact way. As Toda (1979) notes, "utilities are exogenous variables in the decision theory: the 'rational' reasoning process of the decision theory begins when all the utilities have been given or measured, asking no question why the utility of a certain final state should be such-and-such nor whether all the utilities given are consistent as a value system. This modesty of the decision theory is rather puzzling, considering that it has Bayesian theory for subjective probability". I will return to this point toward the end of my essay where I will discuss the formal concept of rationality used by decision theory.

In view of this, it is not surprising that there appears to be a tendency in present-day psychology to start with conflicting values which belong to the individual's motivation in describing decision. Thus in a recent paper by Nowakowska (1979) we find 'economic' value and value of success (which depends negatively upon the subjective probability of success) mentioned as motivational variables. These two values are then combined to give an overall expectation, to be maximized by the individual. This is, of course, a solution to conflict of values. However, what is characteristic of such a solution is that it is related to a specific type of conflict, As far as I know, attempts to provide a general criterion to solve *any* type of conflict of values have been given up, presumably as hopeless. Also, there is at least one personal parameter in the way values are aggregated, e.g., in the case just quoted, the strength of the desire to achieve success. While it is, again, less satisfactory from the point of view of generality, it certainly is more realistic as a partial answer to the question about how value-conflicts are actually solved.

4. Formally similar problem concerns the social choice which is ethically satisfactory. During the last thirty years an enormous literature has grown around this problem, and is still growing. In the classical version (Arrow 1951), numerical utilities were replaced by preference orderings in order to eliminate the temptation to assume interpersonal exchangeability of utilities, leading directly to Bentham's arithmetic. The subject is very well known, at least in its general outline (for a clear overview, see Sen 1969). It has been, and still is, governed by ethical considerations that are quite elementary but nevertheless indispensable; the absence of a dictator, non-negative influence of individual wishes on social choice, some sort of symmetry in the treatment of individuals and so one.

5. Symmetry is essential to attempts to formalize the concept of distributive justice. Since this topic is comparatively less well-known than the topics previously discussed, I will describe in more detail, using for simplicity's sake a numerically specific case, one possible approach to it.

Suppose three objects, *A*, *B*, *C*, are to be distributed among three persons, each person getting exactly one object. There are, of course 3! ways, call them divisions, of doing this. We assume that the profile of preference orderings of the objects is given. By the *distribution rule* I mean a function that associates with each possible profile of preference orderings a probability distribution over the set of all divisions. (Probability is introduced in order to 'smooth out' the way the objects are distributed, facilitating symmetric treatment of people's conflicting wishes; the paradigm is tossing a coin for it.) Which distribution rules are ethically satisfactory for representing distributive justice?

It turns out that even in such a simple case there are at least two possible answers. A 'stronger' concept of justice would demand that all individuals have the same probability of getting the object of their *k*-th choice ($k = 1, 2, 3$). Unfortunately a distribution rule satisfying this requirement may easily be Pareto non-optimal, in the sense that it leads with non-zero probability to divisions that are dominated by some other divisions. An alternative conception would give everybody the same probability of being the *k*-th person to choose. Such a rule automatically satisfies Pareto optimality, but it has the drawback that for some profiles of preference orderings it leads to very asymmetric probability distribution; people with somewhat eccentric tastes have a much higher probability of having their wishes satisfied (Szaniawski 1979).

The practical significance of such an analysis is obviously limited since probabilistic decisions in social life often are difficult to effect. Also, there seems to be some reluctance to use them as a means of solving problems. Philosophically, this decision-theoretic model seems to throw some light on various intuitions that govern the moral evaluation of possible distribution of goods.

A decision-theoretic analysis of the concept of justice has recently opened another field of problems. The much discussed theory of J. Rawls finds a natural expression in terms of the familiar utility matrix. Rows are interpreted as alternative social institutions, columns as social classes. According to Rawls, justice demands that a member of society should opt for a social institution *as if he did not know which class he belonged to*. The analogy to decision-making under uncertainty is obvious. A slightly modified version of maximin is one way of solving the problem. If we admit probability of belonging to a given class (e.g., proportional to the estimated size of that class) then maximization of average utility is another possible principle of just decisions. This analysis (Ellsworth 1978), although presently inconclusive, is a valuable formalization of an ethical position, namely of Rawls's theory of justice.

Recent work by Rosenkrantz (1978) is more closely related to actual social practice, since it explicitly tries to answer the question "What is a fair division of burdens and benefits attaching to a cooperative endavor?". The solution takes into account, among other things, the incentive-value of the attributed benefits. The effect on the overall output is one of the factors that determine the solution.

6. A related problem is that of mediation schemes in situations of partial conflict. Such a scheme would have to be ethically satisfactory in the sense of symmetric treatment of individuals; at the same time, it would have to reflect the individual's unequal strategic positions or the scheme would not be acceptable to the interested parties. In other words, a mediation scheme will have to be a compromise between morally just and workable one.

There was a time when this problem aroused considerable interest. It was thought that here was an answer to some of the classical problems of ethics. Of interest here is the book by Braithwaite (1955), *Theory of Games as a Tool for the Moral Philosopher*; this book advocated a rather ingenious method of arriving at a probabilistic compromise in a two-person non-zero-sum game. It is difficult to judge whether such hopes proved to be justified. Work on this problem continues. A recent example is that of Barry (1979) who has re-examined Braithwaite's formulation of the problem and has attempted to solve it using a different principle.

Epistemology

The main epistemological problem that has been affected by decision theory, of course, is that of induction. Several factors have contributed to the effect that discussions concerning inductive inference have gone beyond the age-old platitudes of the so-called enumerative induction and beyond the methods of J.S. Mill which are, in fact, deductive rules. Probability theory and its applications were influential here, mathematical statistics in particular. Its use in the experimental sciences proved that there is much more to inductive inference than had been thought until the last decades of the 19th century. The so-called statistical inference provided a new paradigm of induction. Decision theory entered the area when attempts were made to systematize various forms of induction and to justify their use. First, it was realized that to reach a conclusion is to make a decision, hence the concept of decision function (due to A. Wald) is practically equivalent to that of rule of inference. Second, values were explicitly introduced since the description of any decision is incomplete unless evaluation of its possible outcomes is given. To try to sketch the history of this subject here is neither possible nor necessary. I shall concentrate, instead, on what I consider to be the main issue, i.e., the concept of induction itself.

More traditional view interprets induction as a kind of inference, i.e., a process leading from acceptance of one statement to acceptance of another statement. A decision-theoretic approach demands that the transition be governed by specific values, the so-called epistemic utilities. Epistemic utilities represent, in most cases, a compromise between two conflicting purposes; the search for truth and the search for information. Exact nature of the compromise is a matter of individual disposition, usually the degree to which one is inclined to gamble (Levi 1967; Szaniawski 1976).

Contrary view denies the existence of inductive inference. Instead what we have is transition form one probability distribution to another probability distribution. This represents the actual use that is being made of empirical data. According to this view, acceptance of a statement is a behaviorally meaningless concept, while a probability distribution is immediately applicable to *any* decision problem for which the truth of the statement is relevant.

The degree to which the above two views are really opposed is an open question. Interpreting acceptance as a probability suitably enough appears to be blocked by the lottery paradox. *If*, however, the concept of acceptance is admissible at all, then arises the problem of how to choose one of the competing hypotheses for acceptance. This is a decision-theoretic version of the traditional question "When should one be satisfied (provisionally, at least) with the result of induction?".

The problem is meaningful for an adherent of the Bayesian approach what is shown by the following suggestion by Jeffrey (1975: 114): "Empirical testing starts after a certain amount of would-be discovery has taken place: one or more hypotheses H, H' . . . have been recognized as 'starters' in the race, *i.e* as worth testing. These will be incompatible hypotheses whose denials are non-starters. The management is unable to copy with photo-finishes: a winner can be declared only *if it is far and away the most probable* of the starters after the empirical test has been made" (my italics – K.S.). In order to formalize thise suggestion, let H_1, ... H_N be the competing hypotheses. Let p_i denote the posterior probability of H_i, given the empirical evidence intended as the basis for choice. Then the Bayesian approach takes the form of the following rule: Given the evidence, the hypothesis H_r is accepted iff

$$p_r \geq A \cdot p_s \quad \text{for all } s \neq r.$$

The parameter A satisfies $A > 1$ and is chosen arbitrarily.

Some properties of the A-rule of acceptance are easily obtained. Thus, for instance, if H_r is accepted by the A-rule then its posterior probability exceeds a threshold, defined in terms of A and the number of hypotheses N:

$$p_r \geq \frac{a}{A + N - 1}$$

Also it is seen that the A-rule brings the entropy of the system of hypotheses below a certain level, depending upon A and N. Hence induction, thus defined, is an information-providing process in the classical sense of the word 'information'.

If no hypothesis satisfies the acceptance criterion, then the evidence is considered inconclusive and further tests must be made in order to decide between the hypotheses. The decision-theoretic approach thus helps us to answer the question: when is the existing evidence satisfactory for inductive acceptance of a hypothesis.

Praxiology

The general theory of efficient action, praxiology, is perhaps most naturally influenced by decision theory. Actual relationships between these two disciplines, however, are not so simple because of differences in origin and in historical development. This subject has been discussed in detail by Gasparski and Szaniawski (1977). Since I do not want to repeat the content of this paper, I focus here only on one specific problem, namely on the contribution of decision theory to the analysis of the concept of rationality. This concept which is certainly central to praxiology; according to Oscar Lange, praxiology is nothing but the theory of rational action. Decision theory has done much to elucidate, in more ways than one, what it means to choose rationally under conditions of imperfect knowledge, more precisely, in a special class of such conditions. Let me stress that this is rationality in a doubly limited sense.

First, it must leave some freedom of choice to the decision-maker since, by definition of imperfect knowledge, a unique solution cannot, in general, exist. This is the element of arbitrariness I mentioned in an earlier part of this essay. Freedom of choice may assume different forms: subjective estimates of chance, attitudes toward risk as a factor co-determining valuations, personal parameters of a similar type (such as the Hurwicz index of pessimism), and so on. The concept of rationality becomes effective outside the field of such individual characteristics.

Second, the decision-theoretic concept of rationality is formal. This means that in regard to beliefs it is concerned with consistency only. And in regard to goals it assumes them as given and not subject to further scrutiny. This last point has been described by Blegvad (1978) as follows: "If we accept rationality as something relative to a given goal, we use the concept of what has been called formal rationality. If, on the other hand, we set up claims regarding the goal, the concept used is that of substantive rationality". Clearly, decision-theoretic analysis does not apply to substantive rationality. Its results, therefore, are quite inapplicable in situations in which we are accustomed to evaluate human actions and their goals. For example, is it rational to do oneself serious bodily harm in order to obtain sick leave? Or, is it rational to smoke cigarettes? A person inclined to answer such questions in the negative, without asking for specification of the individual's hierarchy of values, uses a concept of rationality that is quite foreign to decision theory, at least the classical version of decision theory. The same goes for attempts to draw the line between prejudice and sound belief in the absolute senses of the words.

In view of the limited scope of decision-theoretic analysis, it is perhaps a consolation for a philosopher that so much is still left open in the semantics of rationality, in spite of all the (welcome?) help from decision theory.

What about feedback to the theory of decision-making? How has the theory profited by having such close ties with philosophy? It ought to be apparent by now that quite a number of formal developments were motivated by the desire to answer purely philosophical questions. To enumerate such cases would be tedious since it would imply going once again over the ground covered by the above survey.

* Translated from the Polish manuscript by the editors.

REFERENCES

Arrow, K.J., 1951. *Social choice and individual values*, J. Wiley, New York.

Barry, B., 1979. "Don't shoot the trumpeter. He's doing his best! Reflections on the problem of fair division.", *Theory and Decision* 11 (2).

Blegvad, M., 1978. "The content and status of rationality assumptions in social science.", *Danish Yearbook of Philosophy* 15.

Braithwaite, R.B., 1955. *Theory of Games as a Tool of a Moral Philosopher*. Cambridge: Cambridge University Press.

Domotor, Z., 1978. "Axiomatization of Jeffrey utilities.", *Synthese* 39 (2).

Ellsworth, L., 1978. "Decision-Theoretic Analysis of Rawls's Original Position". In: C.A. Hooker, J.J. Leach and F.F. McClennen (eds.), *Foundations and Applications of Decision theory*. Dordrecht-Boston: Reidel Publishing Co.

Gasparski, W. and K. Szaniawski, 1977. "Praxiology and Decision Theory." In: H. Jungermann and G. de Zeeuw (eds.), *Decision Making and Change in Human Affairs*. Dordrecht-Boston: Reidel Publishing Co.

Jeffrey, R., 1965. *The Logic of Decision.* New York: McGraw-Hill.

Jeffrey, R., 1978. "Critique of the Popper Program.", *Synthese* 30 (1,2).

Jeffrey, R. 1978. "Axiomatizing the Logic of Decision", In: C.A.Hooker, J.J.Leach and E.F, McClennen (eds.) *Foundations and Applications of Decision Theory.* Dordrecht-Boston: Reidel Publishing Co.

Kotarbiński, T., 1955. *Traktat o dobrej robocie* (in Polish). Wrocław, Warszawa, Kraków: Ossolineum. *Praxiology: An Introduction to the Science of Efficient Action.* Oxford: Pergamon Press 1965 (English edition).

Levi, I., 1967. *Gambling with Truth. An Essay in Induction and the Aims of Science.* New York – London: A. Knopf – Routledge and Kegan Paul.

Neuman, J. von and O. Morgenstern, 1944. *Theory of Games and Economic Behaviour.* Princeton: Princeton University Press.

Nowakowska, M., 1979. "A New Model of Decision Under Risk", Abstract from the Seventh Research Conference on Subjective Probability, Utility and Decision-Making, Göteborg. Sweden, August 27–31, 1979.

Rosenkrantz, R.D., 1978. "Distributive Justice". In: C.A. Hooker, J.J. Leach and E.F. McClennen (eds.), *Foundations and Applications of Decision Theory.* Dordrecht-Boston: Reidel Publishing Co.

Savage, I.J., 1954. *The Foundations of Statistics.* New York: Wiley.

Sen, A.K., *Collective Choice and Social Welfare.* San Francisco: Holden-Day.

Sjöberg, L., 1979. "Life Situations and Episodes as a Basis for Situational Influence on Action.", paper read at the Symposium on the Situation in Psychological Theory and Research, Stockholm, Sweden, June 19–22, 1979.

Suppes, P., 1973. The Concept of Obligation in the Context of Decision Theory. In: J. Leach, R. Butts and G. Pearce (eds.). *Science, Decision and Value.* Dordrecht-Boston: Reidel Publishing Co.

Szaniawski, K., 1976. "Types of Information and Their Role in the Methodology of Science", In: M. Przełecki, K. Szaniawski and R. Wójcicki (eds.), *Formal Methods in the Methodology of Empirical Sciences.* Wrocław: Ossolineum [see this volume, pp. 167–176.].

Szaniawski, K., 1979. "On Formal Aspects of Distributive Justice". In: E. Saarinen, R. Hilpinen and I. Niiniluoto (eds.), *Essays in Honour of Jaakko Hintikka.* Dordrecht-Boston: Reidel Publishing Co. [see this volume, pp. 200–209.].

Toda, M., 1979. "Emotion and Decision-Making" (Paper delivered at the Seventh Research Conference on Subject Probability, Utility, and Decision Making, Göteborg, Sweden, Autgust 27–31, 1979).

Von Winterfeldt, M., 1980. "Structured Decision Problems" (unpublished).

24
THE CONCEPT OF UNRELIABLE INFORMATION

The expression 'unreliable information' presupposes some sort of semantics, since it refers to the possibility of a discrepancy between what the information says and the actual state of affairs. Now, information *says* something, in a non-derived sense of the word, only if a correspondence has been established between its sign-aspect and the state of the world it points to. This already constitutes rudimentary semantics.

On the other hand, the widespread use of the word 'information' presupposes nothing of the kind. It is based exclusively on a probabilistic relation between possible values information can assume and 'states of the world'. The value of information has thus no interpretation to start with. Nevertheless, it is currently said that there are instances of unreliable information. The issue is further confused by the well-known ambiguities of the word 'information' itself.

Let us start with a few elementary distinctions. Generally speaking, the concept of information is applicable when it is known what *carries* information and what the information *is about*. Let X and Y be two sets; for simplicity we assume them to be finite. X is interpreted as the set of possible states of the world (the object of information). Y is interpreted as a set of possible signals, or messages, about X (the carriers of information). Let $0y$ stand for 'y is observed' and Ax stand for 'x is the actual state of the world'. It is assumed that, at any given moment, there is exactly one such $x(y)$ that $Ax(0y)$. The information provided by any y in Y consists, broadly speaking, in a reduction of uncertainty concerning: which x is such that Ax? For this purpose, some connection must be assumed between elements of X and those of Y.

One well-known way to define this connection is in terms of probability. Let $p(x, y)$ be joint probability distribution function of the random variables x and y, having X and Y as their sets of values, respectively. The marginal and conditional distribution functions will be denoted by $p(x)$, $p(y)$, $p(x \mid y)$, $p(y \mid x)$. Then the expected reduction of uncertainty about x, due to observation y in Y, is usually defined by the Shannon formula,

$$I(x, y) = E_{p(y)} H(x \mid Y = y), \tag{1}$$

i.e. the entropy of x, minus the conditional entropy of x given y, averaged over all y in Y. [This is not the only way of defining the decrease of uncertainty; see Szaniawski (1974).]

The triple $I_p = \langle X, Y, p \rangle$ may be called *information structure*, in order to distinguish it from *specific information*, i.e. any element of Y. Another way to look at is to treat I_p as an (indirect) question concerning the actual state of the world; any y in Y would then be an answer to that question.

The connection between X and Y could also be based on some kind of convention, or code, by virtue of which a y denotes the reduction of uncertainty from the whole set X to its subset X_y. In other words, y would mean that the actual state of the world belongs to $X_y \subseteq X$. Let, therefore, s be such a mapping of Y into 2^X, subject to the condition that

$$\bigcup_{y \in Y} X_y = X. \tag{2}$$

A special case is obtained by postulating that the sets X_y be pairwise disjoint, i.e., that $\{X_y\}_{y \in Y}$ be a partition of X. Further specialization consists in assuming that s is a one-one correspondence between the elements of X and those of Y. Under this assumption, any y in Y completely removes the uncertainty, by pointing out the actual state of the world (whether correctly or not, is an open question). Let us designate by y_x this y which, by the convention s, points out x as the actual state of the world. From now on, we shall limit our attention to this last case, which simplifies the notation and conceptually is quite sufficient. It follows that we assume the sets X and Y to be equipotent (since a one-one relation s between their elements can be defined).

We thus have two types of information structure: $I_p = \langle X, Y, p \rangle$ and $I_s = \langle X, y, s \rangle$, depending on the type of connection between elements of the sets X and Y. In the standard usage, the concept of reliability is applicable to information structure *as a whole*. In other words, it is the source of information that may be called reliable (or have a given degree of reliability), rather than any specific information it generates. Only in a derived sense can we speak of specific information as 'reliable', meaning by this that it belongs to a reliable information structure.

In the case of I_s, reliability seems to be a qualitative concept. The structure I_s will be called reliable if it generates true information only, i.e. if it satisfies the condition:

$$\text{for any } x, \qquad Oy_x \Rightarrow Ax. \tag{3}$$

Otherwise, I_s will be called unreliable.

Condition (3) is a strong desideratum and we may wonder how often an information structure is treated as reliable in this sense. The answer is, however, less nihilistic than might be expected. An information structure is quite often treated as reliable, although the possibility of an incorrect answer is, in principle, not excluded. We thus have implicit trust in an enormous amount of verbal information, obtained from handbooks, encyclopaedias, expert opinions etc., on any topic outside the sphere of our own competence.

Whether to base a practical decision on information which is not literally reliable in the above sense, depends of course on the nature of the decision. Unless, however, the decision has very serious consequences, we are satisfied to assume that the actual state of the world is the one designated by the verbal information in 'practically reliable' structure. To decide on such an 'as if' basis is an effort-saving procedure.

The information structure I_p makes it possible to speak of reliability in quantitative terms. Suppose that y has been observed. Then the probability, say $r(y)$, of correct identification of the state of the world is equal to $\mathrm{Max}_x\, p(x \mid y)$. The quantity $r(y)$ may be interpreted as the degree of reliability of I_p, conditional upon y. In order to make it independent of the variable y, and thus an absolute property of I_p, some reduction to a single value must be effected. One natural way to do this is by means of the minimum:

$$r \overset{df}{=} \underset{y}{\mathrm{Min}}\ \underset{x}{\mathrm{Max}}\ p(x \mid y). \tag{4}$$

The degree r of reliability of I_p is thus defined as a minimal probability of correctly identifying the state of the world on the basis of Y. Alternatively, though perhaps less intuitively, we could define r as average of $r(y)$ over all y in Y.

The lowest degree of reliability in the sense of def. (4) is, of course, equal to $1/n$, where n is the number of elements in X. It characterizes I_p when at least one y makes all states of the world equiprobable. Maximal reliability, i.e. $r = 1$, occurs when each y solves the problem completely:

$$\bigwedge_y \bigvee_x p(x \mid y) = 1. \tag{5}$$

Condition (5) establishes a one-one correspondence between X and Y (under the assumption that no state of the world is impossible). If this correspondence coincides with the code s, then s satisfies condition (3). In other words, a maximal degree of reliability is simply reliability if the semantic interpretation of each y in Y accords with the maximum conditional probability:

$$s(y) = x \quad \text{iff } p(x \mid y) = \underset{x}{\mathrm{Max}}\ p(x \mid y). \tag{6}$$

The maximum in (6) is equal to 1 if condition (5) holds. An approximation to 1 is, in practice, more realistic. Uniqueness of the maximum is guaranteed under the relatively weak condition that it is greater than one-half.

Condition (6) seem reasonable as a postulate on s. If it is violated by some y_x then that y_x is misleading as an indication of that x. Misleading not in the sense that the possibility of $0y_x \wedge \sim Ax$ is not excluded (this would be too much to ask), but in the sense that some state other than the x state of the world is more probable, given y_x, than x. Misinformation of that type is not infrequent (sometimes even intentional); it ought, however, be distinguished from genuine unreliable information.

To sum up these brief remarks. In the probabilistic information structure $I_p = \langle X, Y, p \rangle$, the information carrying elements of Y are uninterpreted: they may be events of any kind. For this reason, they cannot be described in terms of their truth value. The (semantically interpreted) correspondence between elements of Y and those of X can either be introduced independently in the form of a code, or it can be derived from the probability structure.

In the first case, reliability of information structure cosists in conformity of any possible y with the state it denotes by virtue of the code s. Using the word 'truth' in a somewhat loose way, we could say that I_s is reliable if any specific information it generates is true. If, additionally, the probability relation between X and Y is given, then the code s (unless it is misleading) ought to satisfy the condition (6) of consistency with p. The reliability of I_s is then equivalent to the maximum degree of of reliability of I_p.

In the second case, the code s is *defined* in terms of the probabilities, by means of the same condition (6). Thus, for instance, positive result of a medical test is interpreted as information of a certain illness, because the probability of the latter given this result is higher than the probability of non-illness. Being derived from the probability structure, the code s is automatically consistent with it.

It ought to be noted that for the purpose of decision-making the interpretation of the y values is irrelevant, since the decision is made to depend directly on the observed y, circumventing any beliefs we might have about the elements of X. Therefore, as long as the probability structure is given, the relation s, hence also reliability in the sense of (3), is superfluous. It is, however, easy to notice that I_s structures are, in fact, widely used and treated as reliable. One possible explanation is that they are much easier to store and make use of than the respective probability structures.

REFERENCES

Szaniawski, K. 'Two Concepts of Information', in: *Theory and Decision*, 5, 1974, pp. 9–21 [see this volume, pp. 157–166.].

25
ON DEFINING INFORMATION

So many definitions of information are known that any new attempt to define new concept calls for justification. The purpose of the present effort can be stated as follows. The standard way to define information is in terms of probability. In this way one aspect of the concept finds the satisfactory expression, viz. unexpectedness: the more unexpected (the less probable) an event is, the more information it carries. Also, since probability itself is a measure, it makes it possible to measure information; this is generally considered an improvement of qualitative concept.

On the other hand, the probabilistic concept of information is inapplicable in a large class of typical uses of the word. Thus, for instance, it would be artificial to define in terms of probability information contained in encyclopaedias, textbooks, etc. What probability to ascribe to the statement that Julius Caesar died on 15th of March, 44 B.C.?

There are also other problems. If it is legitimate to speak of 'true' or 'false' information, then some rudimentary semantics is presupposed. For this purpose, however, mere probability is not enough. Likewise, to predicate 'reliability' of an information source, e.g. a person, implies existence of a previous established relation between the information conveyed by the source and the state of the world it points to. A purely probabilistic definition of information excludes such a possibility.

The above considerations suggest, that, to start with, probability-based information ought to be separated from the information based on some kind of code. Only after the two are well defined, any combination of them can be fruitfully discussed.

Informally, one might say that information consists in identifying, completely or partly, an element of a set, say X, of states of the world. It is assumed, of course, that exactly one x in X is actually the case. Symbolically Ax will stand for: x is the actual state of the world. The set X has more than one element: otherwise the problem of identification does not arise.

The set X constitutes *what information is about*. There will also have to be (potential) *carriers of information*. Let the set of them be denoted by Y. In order to perform the identification (whether complete or not), one element of Y must be singled out, thus becoming an actual information. This 'singling out' may consist in asserting a sentence y out of the set Y of sentences; e.g. in observing an outcome y of an experiment, out of the set Y of all possible outcomes, etc. I shall denote this fact by Oy. It is, of course, postulated that exactly one y in Y is such that Oy. The only y satisfying OY is the actual *information about* X, provided some connection between elements of Y and those of X exists, making y *relevant* to the problem of identification in X. In accordance with opening remarks, I shall distinguish two cases.

First, it may be the case that a y in Y points to a subset of X, indicating that the actual state of the world belongs to this subset. In other words, there is a function s from Y to the power set of X, such that Oy is interpreted as reducing the set X of all possible states of the world to $s(y)$. The information y is genuine, in the sense that it reduces uncertainty concerning the actual state of the world, if the subset of X it points to is neither empty nor maximal. Also, it is natural to postulate that at least one y in Y is correct, in the sense that it points to a subset containing the actual x. The function s must, therefore, satisfy the conditions:

$$\text{For all } y \text{ in } Y\colon \varnothing \neq s(y) \subset X; \ \bigcup_{y \in Y} s(y) = X \tag{0}$$

The identification of the actual state of the world is complete iff the function s satisfies

$$\text{for all } y \text{ in } Y\colon |s(y)| = 1. \tag{1}$$

We shall assume that no y in Y is redundant, i.e. if $y \neq y'$ then $s(y) \neq s(y')$. Under his assumption, (1) implies that s is a one-one relation.

If we interpret potential information based on the relation s as a statement, it is natural to introduce the notion of truth. For any y in Y, y is true iff

$$\text{for any } x \text{ in } X\colon Oy \ \& \ Ax \Rightarrow x \in s(y). \tag{2}$$

Condition (2) states that if y is asserted and x is the actual state of the world, then x belongs to the subset of X referred to by y.

The nature of the function s, connecting the sets Y and X, is here left unspecified. In most cases, however, s will be established by means of a convention, i.e. a linguistic code. The situation is formally summarized as the triple $\langle X, Y, s\rangle = I_s$. I_s will be called *semantic information structure.* I_s satisfying (1) is called perfect information structure, since it supplies complete identification of the actual state of the world.

According to the second approach, represented by the 'classical' theory of information, the connection between the sets Y and X is defined by means of probability. The joint probability distribution $p(x, y) = \text{Prob}(Ax, Oy)$ is postulated. The distribution $p(x, y)$ generates, of course, the marginal and conditional probability distributions, denoted respectively by $p(x)$, $p(y)$, $p(x \mid y)$, $p(y \mid x)$. The triple $I_p = \langle X, Y, p\rangle$ will be called *probabilistic information structure.* On the basis of I_p, familiar concepts of information theory are defined. Thus, any y in Y generates the conditional entropy of X, given y:

$$H(X \mid y) = E_x[-\log p(x \mid y)] = -\sum_x p(x \mid y) \log p(x \mid y) \tag{3}$$

and the average reduction of uncertainty in X, due to Y, is defined as

$$I(X, Y) = H(X) - E_y[H(X \mid y)], \tag{4}$$

i.e. the entropy in the set X minus the conditional entropy given that y is observed, averaged over all the elements of Y. This is a well-known formula and the quantity $I(X, Y)$ is sometimes called 'the amount of information' or simply 'information' on X, contained Y.

The two sets collapse into one, i.e. $X = Y$, if the problem of identification can be solved directly, e.g. by the observation of the element of X such that Ax. Expression (4) then becomes simply the entropy of X, $H(X)$, and this again is called information, which adds to the terminological confusion.

I suggest that only the element of Y satisfying Oy be called information about X, relative to the given information structure, whether semantic or probabilistic. The proposed convention seems justified if we agree that information implies some kind of assertion: an element of Y informs only if it is observed, asserted, singled out in some way out from all elements of the set Y. On this assumption, it is natural to describe $I(X, Y)$ as the average amount of information X, contained in Y.

Often the two information structures are mixed, hence the possibility of misunderstandings arises because of the typically incomplete reliability of semantic information structure. Full reliability of I_s would mean that, within this structure, any information is true, i.e. that it satisfies condition (2). For sake of comparison of I_s and I_p, I will from now on assume that the identification of the actual x in X is complete in the sense of condition (1). In this case, full reliability of I_s reduces to:

$$\text{for all } y \text{ in } Y\colon Oy = As(y). \tag{5}$$

The reliability so defined is a qualitative property (it allows no gradation) and a very strong one. Actual sources of information, e.g. encyclopedias, people, newspapers, etc. (they are in fact collections of I_s information structures), are seldom reliable in this sense. They may, however, be treated as such for practical purposes, on the condition that they seldom turn out unreliable in the sense of strict reliability. To have implicit trust in approximately reliable information structures is an effort-saving practice.

This 'as if' attitude towards semantic information breaks down when reliability of the information structure is low and/or the consequences of erroneous identification in the set X are serious enough.

To be more specific, suppose that a decision depends on x, in the usual sense of utility u being a function of both x and a, where a is an action chosen out of the set A of all possible actions. Acting on the semantic (and perfect) information y means: to choose this action in A which maximizes $u(a, s(y))$. If condition (5) is not satisfied, the actual state of the world may be different from $s(y)$. In the case of a sufficiently important decision problem, the decision-maker will want to know, before committing himself, what are the chances that the semantic information is false. This is where probabilities comes in.

The most rudimentary way to characterize reliability of semantic information structure is by means of a single index r, interpreted as the probability that the information provided in this structure is true:

$$r = p(s(y) \mid y), \text{ for all } y \text{ in } Y. \tag{6}$$

If r is known the utility obtainable under information y, say $u(y)$, cannot be computed exactly. We can, however, determine its lower bound by assuming the worst case:

$$u(y) \geq r \operatorname*{Max}_a u(a, s(y)) + (1 - r) \cdot \operatorname*{Min}_a u(a_y, x), \tag{7}$$

where a_y is the act chosen under y, i.e. satisfying

$$u(a_y, s(y)) = \operatorname*{Max}_a u(a, s(y)). \tag{8}$$

(I am disregarding here the technical problem of possible of possible non-uniqueness of a_y.) Maximization of the lower bound of utility constitutes a possible criterion of decision-making, and certainly a safer one than simple maximization of $u(a, s(y))$, i.e. the choice of a as if y were true, in spite of the non-validity of condition (5).

Maximization of expected utility, given y, would be still better as a criterion of decision (because it is more precise). For this purpose, however, the crude index of reliability is not enough. What the decision maker needs is the knowledge of probabilities of different types of error, i.e. the conditional probability distribution $p(x \mid y)$, for all y in Y. These conditional distribution makes it possible to calculate the (expected) utility $u(a \mid y)$ of an action in A as

$$u(a \mid y) = \sum_x u(a, x)p(x \mid y). \tag{9}$$

Since complete joint probability distribution $p(x, y)$ implies the conditional probabilities $p(x \mid y)$, it is tempting to strengthen the probabilistic description of the semantic information structure to $p(x, y)$; the more so since the conditional probabilities of this type ('posterior probabilities') are seldom given alone, i.e. without 'prior' probabilities $p(x)$. We then have *both* i_s and I_s, relative to the same sets X and Y. The introduction of I_p is due to decision-making needs.

The two information structures may be of conflicting nature. This happens when a given y points to $s(y)$ by virtue of the code s, while the conditional probability $p(x \mid y)$ is maximized by an x, different from $s(y)$. No doubt, this sometimes is the case. One example: verbal information which is deliberately misleading. The condition of consistency (in this sense) of the two information structures, I_s and I_p, is given by the formula

$$\text{for all } x \text{ and } y\text{: } x = s(y) \quad \text{iff } p(x \mid y) = \operatorname*{Max}_x p(x \mid y). \tag{10}$$

Such consistency is, of course, assured if the reliability r of sematic information structure is not less than 1/2. In such a case, each conditional distribution is unimodal and the state of the world maximizing $p(x|y)$ coincides with $s(y)$. From the point of view of decision-making, this consistency is immaterial. Indeed, the semantic information structure becomes superfluous once the joint probability distribution $p(x, y)$ is given. The decision is then made to depend directly on y and it does not matter what the y asserts, since the acceptance of $s(y)$ as the actual state of the world is not needed anymore. One might say that, for the purpose of the decision-making, the probabilistic information structure makes it possible to disregard the belief concerning what *is* the actual state of the world. This explains the fact that in many theoretical contexts probabilistic information structure has suspended semantic one.

The trouble is that complete probabilistic description of the relation between the elements of Y and those of X is often hard to obtain, unless the probability is assumed to be psychological or personal. Typically, only the conditional probabilities of the type $p(y\,|\,x)$ are given. Thus, for instance, if X is a set of a hypotheses and Y represents the outcomes of an experiment, relevant to the hypotheses in X, a hypothesis determines, probability-wise, the outcomes. The reverse relation (i.e. the posteriori probability distribution on X, given an outcome y) cannot, however, be defined unless the so-called *a priori* probabilities of hypotheses are known; which, most often, they are not.

This is a well-known fact, but it has a bearing on the definition of information. Since there is no reason to exclude the case of incomplete probability relation between the elements of X and Y, the simplest solution is to admit probabilistic information structures weaker than I_p, i.e. defined in terms of conditional probability only. Their definitions are as follows.

$$I_p^{\cdot} = \langle X, Y, \{p(\cdot\,|\,x)\}_{y \in X}\rangle \tag{11}$$

$$I_p^{\cdot\cdot} = \langle X, Y, \{p(\cdot\,|\,y)\}_{y \in Y}\rangle \tag{12}$$

Instead of the joint probability distribution, families of conditional probability distributions appear in the expression above, thus establishing a connection between the carriers of information y and its objects x.

The second information structure $I_p^{\cdot\cdot}$ has been introduced for symmetry's sake rather than for practical importance, since it is not often met with. As mentioned above, posterior probabilities of the states of the two world are typically given only if prior probabilities are known. However, if this is the case, complete $p(x, y)$ is also given and we simply have the information structure I_p.

The problem, how to utilize information provided by the structure $I_p^{\cdot}$, belongs to the classical repertoire of mathematical statistics. It is not the purpose of this paper to discuss it. I have included this case in the present exposition, since any definition of information that would leave out the 'one-way' probabilistic relation from X to Y is hardly realistic.

To summarize this brief account. Using the concept of information structure, current in the literature, I distinguished between the case of semantic (i.e. functional) and probabilistic relation between the elements of X and Y. Any element of Y is potential information about X. It becomes information when it is 'singled out' in some way: observed, asserted, etc. Semantic information may be said to be true if it points to the actual state of the world. In a number of routine cases we behave *as if* the semantic information were true: we believe the information, transmit it, act upon it, etc. This is a convenient practice. It must be, however, abandoned when the consequences of the erroneous identification of the states of the world are serious enough. Probabilistic description of information makes it possible to calculate the expected utility, which provides the basis for a rational decision. The probabilistic information structure I_p is thus introduced. It can, in principle be inconsistent with I_s. This, however, is immaterial, because I_s is no longer needed as a guide in decision-making.

Finally, it has been pointed out that probabilistic relation between elements of Y and those of X is often weaker than complete joint distribution. Typically, for any state of the world, conditional (or parametric) distribution on the information set Y is given. Such 'weaker' probabilistic information structures must, therefore, be allowed. Convinced Bayesians will, of course, disregard this last remark.

26
RATIONALITY AS A VALUE

The proponents of rationality do not have an easy life nowadays, above all because the reality which now surrounds us – I have in mind the one over which in principle we can have influence, i.e. social reality – defies the requirements of rationality.

Of course, in answer to this problem one can transfer rationality from the world of facts to the world of postulates (where, after all, it always to a certain degree inhered). One can even maintain that the less rationality there is in the real world, the more important it becomes as a postulate. In such a case, however, one would have to recognize it as a value, as something towards which one must strive – and this is not so obvious, at least not without certain reservations. The axiological status of rationality requires explanation both by providing an answer to the question, in what sense rationality is a value, as well as by relating it to other values.

Let us add that also in the area of theory, in philosophical thought, rationality has run into misfortunes. To be sure, not in the sense that it has been negated outright (though this sometimes happens), but in the sense of ignoring or even violating its requirements. Rationality is often given lip service. While recognizing rationality as a slogan, a philosophy is cultivated which in essence is its contradiction, for the requirements of rationality are not respected – especially those formulated by science. Philosophy then approaches art with its irreducible metaphoricalness, while the term 'scientistic' takes on a decidedly pejorative tinge with respect to philosophy. One can value science as a means for the attainment of goals, while simultaneously rejecting its spirit; one can stress with approval the irrational elements in human motivation and appeal to them in a language freed from discipline, etc. The extreme form of this attitude, vulgarized and widespread, is socially dangerous. One can easily find confirmation for this in recent history.

In these circumstances, a defence of rationality in its classical meaning is perhaps justified. To be sure, not defence at any price and in any case without closing one's eyes to the inadequacies of the rationalist idea, those features of the strictly rationalist attitude which make us feel that something is lacking in it.

The main question which I intend to take up here was announced in the title, namely, the axiological status of rationality. The question is *in what sense* rationality is a value and *what kind of* value it is, rather than – whether it is a value *at all.* I assume (perhaps too optimistically?) that in our civilization rationality is on the whole valued. Thus the following questions require explanation. What sort of rationality are we speaking about? Is it only an instrumental value or also an autonomous one? What is its relationship to other values? How does it relate to the world of human emotions?

I do not pretend to give instructions on these matters. Rather, I wish to formulate certain questions and share my intuitions, without being sure that they are commonly accepted.

One must begin with a definition of terms. First of all, I suggest distinguishing rationality of thought and rationality of action. Quite clearly these are closely interrelated, which was known at least from the times of Sir Francis Bacon, who saw basis of rational behaviour in the knowledge of reality. In other words, rational thinking is an indispensable condition for rational action.

The rationality of thinking (perhaps better: beliefs?) consists of three factors: articulation, logical consistency, and empirical verification. Let us examine them in turn.

If there is something sound in the abbreviated statement that a 'rational proposition' amounts to: a 'proposition adequately substantiated', then the requirement of *precise articulation* imposes itself as the basic requirement. For one can hardly speak of the genuine substantiation of a view whose meaning is unclear. The connection of articulation with verification is especially evident in those sciences in which there is a consensus as to the accepted ways of empirical verification, that is, in the natural sciences. Precise formulation there is not an especially debatable problem, for it is, as it were, compelled by the method of verification. In this respect, humanities are in a much worse situation, and it is mainly in this area that debates are conducted as to which propositions can be recognized as adequately formulated.

Of course, the requirement of precise articulation participates in the concept of rationality on somewhat different ground than the requirements of consistency and concordance with experience. First, historically it comes later. The Skeptics were already aware of the traps which the language we use sets for us; however, the requirement of precision in the form of an expanded and more adequately based knowledge of semantics did not appear until the present century, when a particulary high rank was accorded to it by neopositivism and related doctrines. The well-known attempt was then made (in the interwar years) to draw the line between meaningful statements, that is, ones formulated adequately with respect to their verifiability, and meaningless statements which can neither be confirmed nor rejected. Today we know that this was an unsuccessful attempt, based on naive assumptions concerning the relation between language and the reality which it describes. The motivation for this effort has not been invalidated, however.

Another peculiarity of the requirement of precise formulation is that it itself has hitherto defied all attempts to give it a precise form. Hence its controversiality. Inclusion of this requirement in the concept of rationality will not be universally accepted. And even if there is an agreement as to the principle, the boundaries of this requirement will be debatable. On the other hand, precision of thought is, I believe, an essential component of rationality. The consequences of violating its requirements are particularly felt in social life. They are partly responsible for many illusions, for the effectiveness of many dishonest efforts at persuasion. During the interwar period some popularity was gained by a doctrine advocated by Alfred Korzybski, according to which the majority of social misfortunes stem from semantically poor language, from disturbances in communication, between people. Thus semantic knowledge would be something like a panacea for social ills of our world. Though it is difficult to share this point of view, it is nevertheless true that to a certain degree we are slaves of the language which we use. Awareness of this fact is the first step toward liberation. Quite another thing – and much more difficult – is attentiveness to precision in formulations which claim to express our rational convictions.

The requirements of *consistency* appear to be least in need of defence. Contradiction is universally experienced as a violation of rationality; the appearance of contradiction sets defence mechanism into motion to eliminate it or to negate its existence (the latter case occurs most often when the elimination of a contradictions would require one to give up a belief to which one is attached). In any case, there is no doubt that respect for the requirements of logic is the oldest and least questioned component of rationality. It is also the best known, owing to the hight level of development of the science of logic attained in the last century.

Today this requirement should be expanded so that, beside logic in the narrow sense of the term, it would also include probability theory. For just as logic imposes conditions of consistency on statements which we recognize as true, the calculus of probability imposes conditions of consistency statements of whose truth we are only partially convinced, with various degrees of intensity.

The third element of the rationality of beliefs, i.e. the requirement of adequate *empirical validation* causes no fewer difficulties than the first one (that is, the requirement of precise articulation). It is quite natural that in this question we look for a model in science, and especially in those fields which supply the most adequately validated assertions. However, the interpretation of scientific procedures encounters difficulties which are all too well known to be worth taking up here. Science is conscious of its methods to a small degree only; it suffices that it has mastered these methods in practice. Hence it is necessary to undertake the task of abstracting these methods from research practice. Owing to the level of complexity of modern science, this is a difficult task, subject to the risk of error. It is not surprising that debates on the subject of methods of validation accepted in science still concern such basic question as e.g. whether science validates its assertions through inductive reasoning.

Since we are still a long way from a convincing codification of scientific procedures, one must recognize than the problem of validation of beliefs in the light of evidence supplied by experience remains open. Debatable is even (and now the subject of much discussion) the question whether experimental data rather validate the *choice* of beliefs from the set of alternative possibilities or the *distribution of intensity* of beliefs over elements of this set.

We can sum up this brief description of the concept of the rationality of beliefs. It turns out that it is not as clearly defined as one might wish. This concerns above all the question of precise articulation (in the language of neo-positivists: meaningfulness) of pronouncements which is unresolved in the form of intersubjectively applicable criteria. This is not because there were no efforts made. Quite the contrary, much effort was made to establish these criteria. In the light of the results, it would appear that the goal is basically unattainable – at least with respect to the language presently used by science which is universally accepted as the paradigm for rationality.

The boundaries of rationality are fuzzy in different places as well. Since there are no precise criteria of correct articulation, then also the application of the literal requirements of logic (in the sense of consistency) to pronouncements formulated in natural language may turn out debatable. The concept of empirical verification is inadequately defined as well – for reasons which we have just discussed. Thus in the final analysis none of the defining components of rational thought has been described by clear and unambiguous criteria.

After stating this fact, let us not fall into nihilism, however. That the world 'bald' is imprecise does not imply that bald men (and also non-bald men) do not exist. This is also the case with unquestioned instances of rationality (or irrationality) in thinking. Moreover, in light of the statement of fact that the boundaries of rationality are imprecise, a significant shift in accents has taken place in recent times. Rationality today has more the status of a guiding principle, a group of general instructions not without a certain ethical tinge than a code of directives whose application would be automatic, as it were. I believe that this is precisely what Karl Popper had in mind when he described the concept of rationality in the second volume of his famous book on the open society. Rationality in his interpretation consists in the readiness to subject all judgements to criticism – one's own and especially those of other people. Hence this is a concept formulated in psycho–sociological rather than in logical categories. The ethos of rationality becomes more important than specific directives.

Let us now examine the concept of the rationality of behaviour. I shall limit myself here to individual behaviour, omitting collective behaviour, as this will suffice for the purposes of this article.

In the traditional interpretation action is rational when it is effective with respect to the goal. This statement contains a great oversimplification. After all, praxiology specifies other qualities of action, e.g. economy, which claims to be a part of the concept of rationality. Moreover, the very concept of effectiveness requires elucidation. I suppose that we would call rational an action which, though not attaining its goal, was carried out according to a method which leads to the goal in a very high percentage of cases. Thus it is not the immediate effect that determines rationality, but the principle by which the subject is guided. However we define effectiveness, the following distinction seems indispensable. Action can be effective in reality or only in the opinion of the acting subject. In the first case, we call it objectively (or really) rational, in the second – subjectively (or methodologically) rational. Both of these variants are useful for a description of human behaviour.

In abbreviated form this is the classical conception of the rationality of action. In contemporary literature of the subject, especially in the one which attempts to construct formal models of human behaviour, a different approach has become widespread. For its starting point it takes the concept of decision, which – as we know – is an act of choice. Choice in turn assumes a well defined set of possible actions, among which one is to be undertaken. Their effects are evaluated in various ways and also depend on the factors over which the subject does not have decisive influence.

Rationality in this case would consist in making a choice which would lead to the effect most highly valued. Owing to complications resulting from the fact that the subject is not fully in control of the situation, different variants of rationality in this interpretation suggest themselves. There is no reason to delve into the differences between them here. It will be sufficient to describe what they have in common: they refer to the situation of choice, and thus they organize both its external as well as internal determinants – in the form of evaluations made by the acting subject. The latter, of course, must meet certain conditions of consistency (e.g. the condition of transitiveness) if they are to be called rational. Rationality of the choice itself consist in turn in maximization (in the meaning agreed on) on the value attributed to the effects of action by the subject who makes the choice.

Thus we have to do with two conceptions of rationality of action. One of them refers to (real or imagined) effectiveness with respect to the goal, the other – to a choice maximizing a subjective value. They are not equivalent concepts, but both of them assume a certain description of reality: concerning causal relations, the possible ways of behaviour in given situation, etc. Hence the rationality of beliefs is in this context logically prior to the rationality of behaviour based on these beliefs.

After this very brief terminological explications, let us return to the question concerning the axiological status of rationality. I believe that one has to take this up separately for rationality of thought and rationality of action, though, as I have already mentioned, they are closely interrelated. And one must do this because the answer is, I believe, not the same in the two cases.

An almost universally accepted thesis – at least in our civilization – is that rationality of thought is an autonomous value, that it is something worth striving for apart from eventual practical benefits which ensue from views meeting the requirements of rationality. This is how we were shaped by our cultural tradition, despite its heterogeneity, in which Cartesian and Kantian thought is only one of elements.

The above statement requires a certain qualification. Namely, agreement to recognize rational thought as an autonomous value can turn out to be only a formal agreement – owing to significant leeway in interpretation of this concept, especially with reference to articulation and verification. In essence we have here to do with a case quite common in philosophy (and not only in philosophy), when using a term with a clear positive meaning, we ascribe to it widely differing meanings. This is the only explanation for the paradoxical fact that rational thought is professed as a value by philosophers whose thinking is diametrically opposed in this respect, e.g. by Hegel and Popper. Equivocation is not always so flagrant (and hence trivially obvious): differences between a more and less rigorous interpretation of the requirements of rational thought are sometimes much more subtle and thus more difficult to identify.

The relatively universal agreement on granting rational thought a place (thought not necessarily the highest place) among absolute values is obviously not an historical accident. It becomes understandable in view of the high status which was given to cognition from the very dawn of our civilization. One can, of course, take the position that there are other ways of cognizing the world than those we find in science. The proponents of this view are not obligated to believe in the autonomous value of rational thought. Even among them, however, there are some who either try (with doubtful results) to reconcile non-rational ways of cognition with the requirements of rationality or recognize two parallel conceptions of cognition and thus respect rational thought in some limited area.

It would be difficult to apply axiological qualification of rational thought to rational action. Having been forced to give up the Enlightenment illusions, today we know all too well that one can rationally strive towards unworthy goals. Since this is so, then there seems to be cogency in the view that rational action is a purely instrumental value and hence becomes positive or negative, depending on the goals which it serves. The quality of the goal would be decisive in the case of equating rationality with effectiveness, whereas with respect to the 'optimization' variant, rationality will depend on the subjective nature of the valuations, which can be noble as well as wicked.

There is a certain potential dissonance in putting the matter this way. We have become accustomed to associating a positive meaning with the word 'rational'. We rather unwillingly accept a statement of fact such as: 'Hoess rationally solved the problem of the mass annihilation of corpses in Auschwitz'. Such types of statements are avoided, thought there is nothing in the concept of rational action itself which should exclude them. Rather than ignoring this dissonance, it perhaps would be better to become aware of it. If rational action can serve anti-values, then in exceptional cases it can also turn against rational thought itself. In essence, this is what happens, as the poet inform us:

From the small seed of truth bring forth the plant of falsehood.
Do not follow those who lie ignoring reality.

The praxiological correctness of this recommendation comes from the fact that falsehood rooted in truth acquires persuasive power, for it creates the illusion that it is not he falsehood. Hence an illustration of this phenomenon: rationally selected means can destroy – on a social scale – rational thought.

In characterizing rational thought, I have tried to stress as strongly as possible the fact that, owing to the lack of absolutely precise objective criteria, within certain limit we are forced to apply a subjective measure. Then factors come into play which are not subject to logical codification. The cognitive ethos then acquires importance, that is: criticism (in Popper's interpretation the central component of rationality), objectivism, the tendency to resolve cognitive dissonance in a way which would not harm truth. Rational thought is definitely connected with these values – namely, to the extent that they participate in determining its ethos.

Since we have been speaking of the relation of rational thought to other values, one should mention that its practice requires traits of character which are universally regarded as valuable. It requires courage, especially civil courage, which is indispensable for adhering to beliefs which are either unorthodox with respect to state-enforced dogmas or are unpopular in one's own circle. It requires the abandonment of illusion, and this is generally very difficult. It is said: 'to look straight in the face of truth.' Apparently such looking is sometimes unpleasant.

Rationality promotes responsibility for one's words – both when more precise articulation of what one has asserted is required as well as when justification is at issue. Through openness to criticism the rationality of beliefs is the opposite of initiation, and in this sense it is democratic. One can see this especially in the case of science, in which each person is equally empowered to control validation.

The practice of rationality in thought is also a matter of character, because this attitude is generally unattractive – in the sense that it requires moderation in expression of judgements, whereas an extremism is more effective and more satisfying emotionally. Exciting radicalism unites people in the feeling of their superiority to other, which for many is a source of satisfaction. For this reason the attainment of rational restraint is a form of triumph over one's self.

The social value of rational thought was raised by many writers. In our century Bertrand Russell devoted nearly all of his popular philosophical work to it. It is the main motif of Popper's work *The Open Society and its Enemies*, written during the last war and directed against the totalitarianism in all its forms.

They and people who think like them are accused of naive faith in a social efficacy of directives contained in the concept of rationality. Indeed, the observations of social life does not inspire us with optimism in this regard. This model of rationality seems to be a rather fragile weapon in the age of highly developed techniques of persuasion using audio-visual methods and reaching millions of viewers and listeners. The motto of such well equipped propaganda could be a fragment full of bitter irony from the already quoted poem by Miłosz, *Children of Europe*:

But for whom they are used against,
Make ambiguous words your weapon,
And bury clear words in the darkness of encyclopedias.
Judge no words before the bureaucrats
Have checked in files who speaks these words.
The voice of passion is better than the voice of reason,
For the dispassionate cannot change history.

However, one can take the position that a philosopher – Russell or Kotarbiński – fulfils his task if in his work he is the spokesman for a threatened value, apart from the scale of influence exerted. If we can agree on this, then the present time should be recognized as a time of the defence of rational thought.

Rationality is often opposed to emotions. According to such an interpretation, they are mutually exclusive attitudes: either one is guided by emotions or by reason. This a troublesome slogan. Its persuasive power comes from trivially correct observation that strong emotions deform our judgement of reality. The most frequently encountered variety of such deformation is wishful thinking.

By defending objectivity of a judgement against the unauthorized influence of desires or repulsion, this slogan serves rationality well. It does this unwisely, however, for it reinforces the stereotype that rationality is the enemy of emotional life, that it requires giving up emotions. We know, however, that the latter are an indispensable element motivating human actions, and thus whoever would demand their elimination would turn man into a computer (programmed by whom?).

The opposition 'reason or feelings' is also used for suspicious purposes. It is appealed to by those who wish to convince us to agree to reality as it is, disqualifying as an irrational fantasy the demand to change the very *conditions* of action. Then one appeals to realism, sometimes perhaps even correctly. However, this realism was used so many times as a camouflage for conservative tendencies that this should be noted at least as a warning against overuse of the slogan of rationality.

The stereotype of the rational human being as something unemotional computer executing his program is just as repulsive as it is false. It would not be so popular, however, if there were not some germ of truth in it to make us ponder. Here we touch on perhaps the most difficult question: the question of the limits of rationality. Even its most ardent supporters sometimes feel the need for liberating themselves from the bonds which it imposes on them, the need for some sort of leeway, even violation of the normally observed code. Is this need so blameworthy that one must suppress it under all circumstances.

In the well-known (and beautiful) film *Zorba the Greek* says to his young American friend: you have only one fault, you lack a bit of madness. From the standpoint of rationality, an absurd charge, but it seems that we understand him (especially in light of action of the film). Of course, there is the temptation to defend departures from rationality by an appeal to rationality itself. For example, it is said that in creative work success is attained by 'obsessed' people who ignore the rational rules of game. Thus is it not perhaps more rational to be somewhat of a madman from time to time? But if the complete suppression of the Dionysian element, which exists in all of us and requires expression, is unwise, then what are the circumstances in which it should be free to do its work? Where are the rationally drawn boundaries of rationality? It is hard not to feel the grotesqueness of this question, and so I shall leave this line of thought without a conclusion.

Translated by Chester Kisiel.

INDEX OF NAMES

MIX
Papier aus verantwortungsvollen Quellen
Paper from responsible sources
FSC® C105338

If you have any concerns about our products,
you can contact us on
ProductSafety@springernature.com

In case Publisher is established outside the EU,
the EU authorized representative is:
Springer Nature Customer Service Center GmbH
Europaplatz 3, 69115 Heidelberg, Germany

Printed by Libri Plureos GmbH
in Hamburg, Germany